Current Intelligence Bulletin 64

Coal Mine Dust Exposures and Associated Health Outcomes

A Review of Information Published Since 1995

DEPARTMENT OF HEALTH AND HUMAN SERVICES
Centers for Disease Control and Prevention
National Institute for Occupational Safety and Health

> This document is in the public domain and may be freely copied or reprinted.

Disclaimer

Mention of any company or product does not constitute endorsement by the National Institute for Occupational Safety and Health (NIOSH). In addition, citations to Web sites external to NIOSH do not constitute NIOSH endorsement of the sponsoring organizations or their programs or products. Furthermore, NIOSH is not responsible for the content of these Web sites. All Web addresses referenced in this document were accessible as of the publication date.

Ordering Information

To receive documents or other information about occupational safety and health topics, contact NIOSH at

   Telephone: 1–800–CDC–INFO (1–800–232–4636)
   TTY: 1–888–232–6348
   E-mail: cdcinfo@cdc.gov

   or visit the NIOSH Web site at www.cdc.gov/niosh.

For a monthly update on news at NIOSH, subscribe to *NIOSH eNews* by visiting www.cdc.gov/niosh/eNews.

DHHS (NIOSH) Publication No. 2011–172

April 2011

SAFER • HEALTHIER • PEOPLE™

Foreword

Since its inception in 1970 the National Institute for Occupational Safety and Health (NIOSH) has extensively investigated and assessed coal miner morbidity and mortality. This history of research encompasses epidemiology; medical surveillance; laboratory-based toxicology, biochemistry, physiology, and pathology; exposure assessment; disease prevention approaches; and methods development. The experience gained in those activities, together with knowledge from external publications and reports, was brought together in 1995 in a major NIOSH review and report of recommendations, entitled *Criteria for a Recommended Standard—Occupational Exposure to Respirable Coal Mine Dust*. This document had the following major recommendations:

1. Exposures to respirable coal mine dust should be limited to 1 mg/m^3 as a time-weighted average concentration for up to a 10 hour day during a 40 hour work week;

2. Exposures to respirable crystalline silica should be limited to 0.05 mg/m^3 as a time-weighted average concentration for up to a 10 hour day during a 40 hour work week;

3. The periodic medical examination for coal miners should include spirometry;

4. Periodic medical examinations should include a standardized respiratory symptom questionnaire;

5. Surface coal miners should be added to and included in the periodic medical monitoring.

This Current Intelligence Bulletin (CIB) updates the information on coal mine dust exposures and associated health effects from 1995 to the present. A principal intent is to determine whether the 1995 recommendations remain valid in the light of the new findings, and whether they need to be updated or supplemented. The report does not deal with issues of sampling and analytical feasibility nor technical feasibility in achieving compliance.

John Howard, MD
Director, National Institute for
  Occupational Safety and Health
Centers for Disease Control and Prevention

Executive Summary

Information relating to occupational pulmonary disease morbidity and mortality of coal miners available up to 1995 was reviewed in the NIOSH publication: *Criteria for a Recommended Standard—Occupational Exposure to Respirable Coal Mine Dust*, or Coal Criteria Document (CCD). This led to the following principal conclusions concerning health effects associated with coal mining:

1. Exposure to coal mine dust causes various pulmonary diseases, including coal workers' pneumoconiosis (CWP) and chronic obstructive pulmonary disease (COPD).
2. Coal miners are also exposed to crystalline silica dust, which causes silicosis, COPD, and other diseases.
3. These lung diseases can bring about impairment, disability and premature death.

This Current Intelligence Bulletin updates the previously published review with respect to findings relevant to the health of U.S. coal miners published since 1995. The main conclusions are:

1. After a long period of declining CWP prevalence, recent surveillance data indicate that the prevalence is rising.
2. Coal miners are developing severe CWP at relatively young ages (<50 years).
3. There is some indication that early development of CWP is being manifested as premature mortality.
4. The above individuals would have been employed all of their working lives in environmental conditions mandated by the 1969 Coal Mine Health and Safety Act.
5. The increase in CWP occurrence appears to be concentrated in hot spots of disease mostly concentrated in the central Appalachian region of southern West Virginia, eastern Kentucky, and western Virginia.
6. The cause of this resurgence in disease is likely multifactorial. Possible explanations include excessive exposure due to increases in coal mine dust levels and duration of exposure (longer working hours), and increases in crystalline silica exposure (see below). As indicated by data on disease prevalence and severity, workers in smaller mines may be at special risk.
7. Given that the more productive seams of coal are being mined out, a transition by the industry to mining thinner coal seams and those with more rock

intrusions is taking place and will likely accelerate in the future. Concomitant with this is the likelihood of increased potential for exposure to crystalline silica, and associated increased risk of silicosis, in coal mining.

The main conclusions drawn from review of the new information are:

1. While findings published since 1995 refine or add further to the understanding of the respiratory health effects of coal mine dust described in the NIOSH CCD, they do not contradict or critically modify the primary conclusions and associated recommendations given there. Rather, the new findings strengthen those conclusions and recommendations.

2. Overall, the evidence and logical basis for recommendations concerning prevention of occupational respiratory disease among coal miners remains essentially unaffected by the newer findings that have emerged since publication of the CCD.

In summary, as recommended by the CCD, every effort needs to be made to reduce exposure to both coal mine dust and to crystalline silica dust. As also recommended in the CCD, the latter task requires establishing a separate compliance standard in order to provide an effective limit to exposure to crystalline silica dust.

Contents

Foreword iii
Executive Summary iv
List of Figures viii
List of Tables x
Abbreviations xi
Glossary xii
Acknowledgments xiv
1 Introduction 1
2 Coal Workers' Pneumoconiosis 11
    2.1 Surveillance 11
    2.2 Epidemiology 14
    2.3 Mortality 18
    2.4 Toxicology 20
    2.5 Risk analysis 21
3 Other Respiratory Disease Outcomes 23
    3.1 Risk Analysis 24
4 Cancer Outcomes 25
5 Dust Exposure Levels, Control, And Compliance 27
    5.1 Dust Exposure Levels 27
    5.2 Dust Exposure Assessment 27
    5.3 Compliance Policy and Procedures 28
6 Surface Coal Mining 29
7 Summary 31
8 References 33

List of Figures

Figure 1. Prevalence of CWP category 1 or greater from the NIOSH Coal Workers' X-ray Program from 1970–1995, by tenure in coal mining.

Figure 2. Trend in reported dust concentrations for continuous miner operators, 1968–87.

Figure 3. Probability that an individual starting with no pneumoconiosis (category 0/0) will be classified as 2/1 or greater after 35 years of exposure to various concentrations of coal mine dust.

Figure 4. Predicted prevalence of PMF among British coal miners after a 35-year working lifetime by mean concentration of respirable coal mine dust.

Figure 5. Prevalence of simple CWP category 1 or greater among U.S. coal miners by estimated cumulative dust exposure and coal rank.

Figure 6. Prercentage of evaluated miners with rapidly progressive coal workers' pneumoconiosis by county.

Figure 7. Percentage of miners in the NIOSH Coal Workers' X-ray Surveillance Program with coal workers' pneumoconiosis (category 1 or greater) from 1970–2009, by tenure in coal mining.

Figure 8. Percentage of miners examined in the NIOSH Coal Workers' X-ray Surveillance Program with progressive massive fibrosis (PMF) from 1970–2009, by tenure in mining.

Figure 9. Observed and predicted prevalences (%) of CWP category 1 or greater by age group and MSHA District.

Figure 10. Hours worked per underground coal miner, 1978–2008.

Figure 11. Tons produced per hour worked at underground coal mines, 1978–2008.

Figure 12. Respirable coal mine dust: Geometric mean exposures by type of mine (UG=underground, SU=surface), MSHA inspector (INSP) and mine operator (OPER) samples.

Figure 13. Ratio of special inspection sample values to preceding operator compliance sample values by mine size.

Figure 14. Age-adjusted death rates (per million) for decedents age ≥ 25 years with coal workers' pneumoconiosis as the underlying cause of death—United States, 1968–2006.

Figure 15. Years of potential life lost (YPLL) before age 65 and mean YPLL per decedent for decedents aged ≥25 years with coal workers' pneumoconiosis as the underlying cause of death—United States, 1968–2006.

Figure 16. Risks at age 58–60 after 35–40 working years of: PMF; category 2 or greater; 993 ml deficit of FEV_1 in nonsmokers; 993 ml deficit of FEV_1 in smokers.

Figure 17. Risks for category 2 silicosis in relation to respirable silica concentration (<2 mg/m^3) averaged over 15 years.

Figure 18. Respirable quartz dust: Geometric mean exposures by type of coal mine.

List of Tables

Table 1. Predicted prevalence of simple CWP and PMF among U.S. or British coal miners at age 58 following exposure to respirable coal mine dust over a 40-year working lifetime.

Table 2. Predicted prevalence of decreased lung function among U.S. or British coal miners at age 58 following exposure to respirable coal mine dust over a 40-year working lifetime.

Table 3. Excess (exposure-attributable) prevalence of simple CWP or PMF among U.S. coal miners at age 65 following exposure to respirable coal mine dust over a 45-year working lifetime.

Table 4. Excess (exposure-attributable) prevalence of decreased lung function among U.S. coal miners at age 65 following exposure to respirable coal mine dust over a 45-year working lifetime.

Abbreviations

CAO	chronic airway obstruction
CCD	coal criteria document, formally *NIOSH Criteria for a Recommended Standard—Occupational Exposure to Respirable Coal Mine Dust*
COPD	chronic obstructive pulmonary disease
CWP	coal workers' pneumoconiosis
CWXSP	Coal Workers' X-ray Surveillance Program
FEV_1	forced expiratory volume in 1 second
ILO	International Labour Office
mg/m^3	milligrams per cubic meter
MSHA	Mine Safety and Health Administration
NIOSH	National Institute for Occupational Safety and Health
PAH	polycyclic aromatic hydrocarbons
PDM	personal continuous dust monitor
PMF	progressive massive fibrosis
REL	recommended exposure limit
YPLL	years of potential life lost

Glossary

Aerodynamic diameter: The diameter of a sphere with a density of 1 g/cm³ and with the same stopping time as the particle. Particles of a given aerodynamic diameter move within the air spaces of the respiratory system identically, regardless of density or shape.

Chronic obstructive pulmonary disease (COPD): Includes chronic bronchitis (inflammation of the lung airways associated with cough and phlegm production), impaired lung function, and emphysema (destruction of the air spaces where gas transfer occurs). COPD is characterized by irreversible (although sometimes variable) obstruction of lung airways, and should be considered in any patient who has dyspnea, chronic cough or sputum, and/or a history of exposure to risk factors for COPD. The diagnosis should be confirmed by spirometry.

Coal rank: A classification of coal based on fixed carbon, volatile matter, and heating value of the coal. Coal rank indicates the progressive geological alteration (coalification) from lignite to anthracite.

Coal workers' pneumoconiosis (CWP): A chronic dust disease of the lung arising from employment in a coal mine. In workers who are or have been exposed to coal mine dust, diagnosis is based on the radiographic classification of the size, shape, profusion, and extent of parenchymal opacities.

Crystalline silica: Silicon dioxide (SiO_2). "Crystalline" refers to the orientation of SiO_2 molecules in a fixed pattern as opposed to a nonperiodic, random molecular arrangement defined as amorphous. The three most common crystalline forms of free silica encountered in general industry are quartz, tridymite, and cristobalite. The predominant form is quartz.

Excess (exposure-attributable) prevalence: The prevalence (cases/population at risk) attributable to workplace dust exposure (in the case of CWP, the prevalence adjusted for radiographic appearances associated with lung aging).

International Labour Office (ILO) classification system: A standardized method for assessing abnormalities related to the pneumoconioses based substantially on comparison of test with reference radiographs. In the system there are 4 categories of simple pneumoconiosis (categories 0, 1, 2, and 3), with 0 implying no definite abnormality.

Progressive massive fibrosis: Coal workers' complicated pneumoconiosis. Diagnosis is based on determination of the presence of large opacities (1 cm or larger) using radiography or the finding of specific lung pathology on biopsy or autopsy.

Quartz: Crystalline silicon dioxide (SiO_2) not chemically combined with other substances and having a distinctive physical structure.

Respirable coal mine dust: That portion of airborne dust in coal mines that is capable of entering the gas-exchange regions of the lungs if inhaled: by convention, a particle-size-selective fraction of the total airborne dust; includes particles with aerodynamic diameters less than approximately 10 μm.

Acknowledgments

Michael Attfield was the primary author of this document, which was prepared in the Division of Respiratory Disease Studies, NIOSH, under the direction of Dr. David Weissman. He acknowledges the following assistance in the preparation of the report:

Division of Respiratory Disease Studies (DRDS)

Janet Hale, Eva Suarthana, Mei Lin Wang

Health Effects Laboratory Division (HELD)

Vincent Castranova, Kimberly Clough Thomas

This Current Intelligence Bulletin has undergone substantial internal and external scientific peer review with subsequent revision. A draft version of the document was published on the NIOSH website for public comment for 60 days, with notification of its availability via the Federal Register. All input, both internal and external, has been considered, addressed, and responded to in preparation of this final version.

Internal NIOSH reviewers who provided critical feedback to the preparation of the document

Eileen Kuempel, Education and Information Division (EID). Ed Thimons, Office of Mine Safety and Health Research (OMSHR), Robert Castellan (DRDS), Ainsley Weston (DRDS), Eileen Storey (DRDS).

External Expert Peer Review Panel

NIOSH expresses appreciation to the following independent, external reviewers for providing insights and comments that contributed to the development of this document and enhanced the final version:

Robert Cohen, M.D., F.C.C.P.
Cook County Health and
  Hospitals System
Chicago, IL

Dennis O'Dell
United Mine Workers of America
Washington, DC

Joseph Lamonica
Bituminous Coal Operators
  Association
Washington, DC

Cecile Rose, M.D.
National Jewish Health
Denver, CO

1 Introduction

The publication of the *NIOSH Criteria for a Recommended Standard—Occupational Exposure to Respirable Coal Mine Dust* or Coal Criteria Document (CCD) in 1995 (1) followed a long period of extensive research activity focused on exposure to coal mine dust and its health effects in coal miners. From this research, substantial information had emerged about the extent and severity of respiratory disease caused by coal mine dust, its quantitative relationship with dust exposure, its pathology and toxicology, environmental patterns of relevant exposures, and methodologies for assessing these variables. In particular, the findings demonstrated that not only was there a considerable burden of coal workers' pneumoconiosis (CWP) in the U.S. and other countries, but that underground coal miners were vulnerable to other lung diseases, notably chronic obstructive pulmonary disease (COPD). The evidence came from extensive and well-planned epidemiologic and laboratory-based investigations undertaken primarily in the U.S., the United Kingdom, and (West) Germany, with supporting information coming from studies in other European countries and Australia.

The available information was thoroughly summarized in the CCD. It showed that, in 1995, CWP was in decline in the U.S., with downward trends in prevalence in all tenure groups (Figure 4–2 of the CCD (1), included here as Figure 1). This decline was consistent with reductions in coal mine dust exposure mandated by the 1969 Coal Mine Health and Safety Act (1969 Coal Mine Act) (CCD Figure 4–1 (1); Figure 2). Despite this decline in disease levels, NIOSH concluded from review of the surveillance data and quantitative risk estimates based on the epidemiologic studies that the current dust exposure regulations for U.S. coal mines were not sufficiently protective. Consequently, it proposed lower dust limits for coal mines, enhanced medical surveillance, and other requirements.

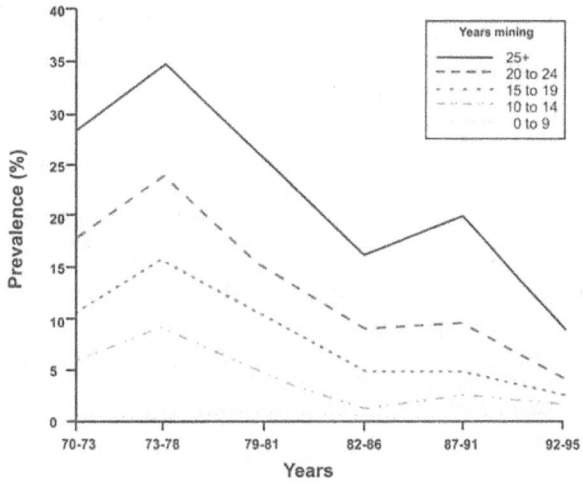

Figure 1. Prevalence of CWP category 1 or greater from the NIOSH Coal Workers' X-ray Program from 1970–1995, by tenure in coal mining. (Figure 4–2 of the CCD (1)).

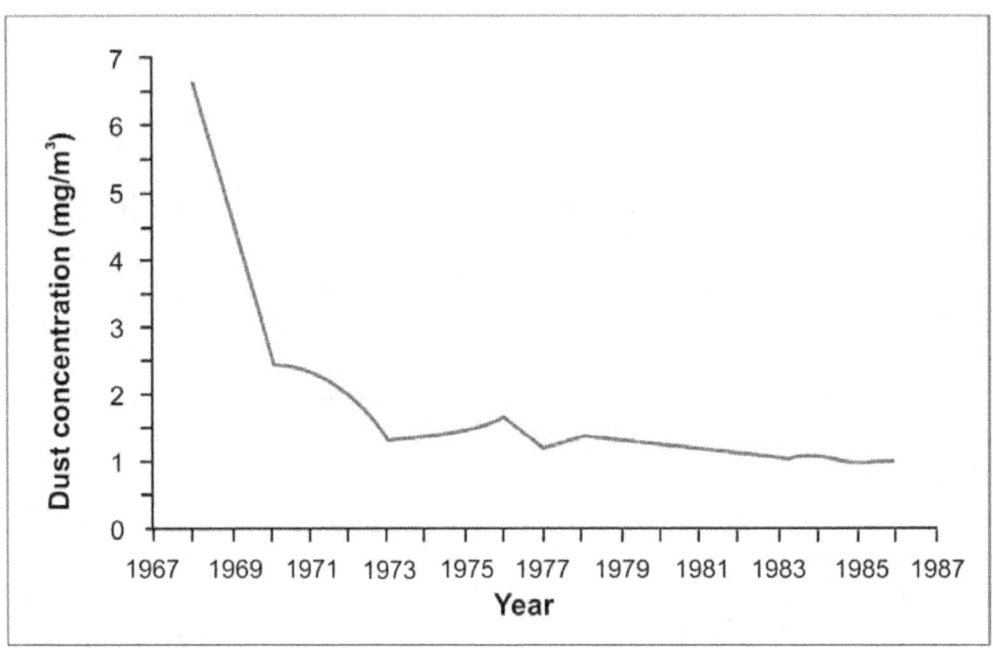

Figure 2. Trend in reported dust concentrations for continuous miner operators, 1968–87. [Figure 4–1 of the CCD (1)] (Source: Attfield and Wagner (2)).

The CCD noted that the current U.S. federal dust limit for underground coal mines dated back to the 1969 Coal Mine Act, which mandated a compliance permissible exposure limit of 2 milligrams per cubic meter (mg/m³) of respirable coal mine dust. This limit was derived from British research, which provided the only quantitative exposure-response relationship available at that time. This exposure-response curve (CCD Figure 7–2 (1); Figure 3) predicted that no cases of CWP as severe as category 2 on the International Labour Office (ILO) classification system (3) would develop among miners who worked 35 years at 2 mg/m³. Similarly at that time, the current information indicated that the disabling form of CWP, progressive massive fibrosis (PMF), was very unlikely to develop from less severe ILO categories (e.g., category 1 CWP). Therefore, adoption of the 2 mg/m³ limit was believed, at that time, to be protective against the risk of disability and premature mortality that accompanies PMF.

Subsequent scientific findings, emerging between 1969 and 1995, disproved some of the assumptions inherent in the adoption of the 2 mg/m³ standard. Firstly, the assumption that miners with CWP less severe than category 2 were at minimal risk of PMF was found to be incorrect. Moreover, additional findings from the British data (CCD Figure 7–6 and Table 4–6 (1); Figure 4 and Table 1), together with new results on U.S. underground coal miners from research undertaken by NIOSH (CCD Figure 7–4 and Table 4–6 (1); Figure 5 and Table 1) showed that there was no threshold at 2 mg/m³ as had been indicated by the original British study (CCD Figure 7–2 (1); Figure 3). Furthermore, the CCD reviewed findings on other lung diseases and their relationship with coal mine dust exposure. It concluded that coal miners were at additional risk of developing ventilatory function deficits, respiratory symptoms, and emphysema in addition to CWP (CCD Table 4–7 (1); Table 2). (Note that the

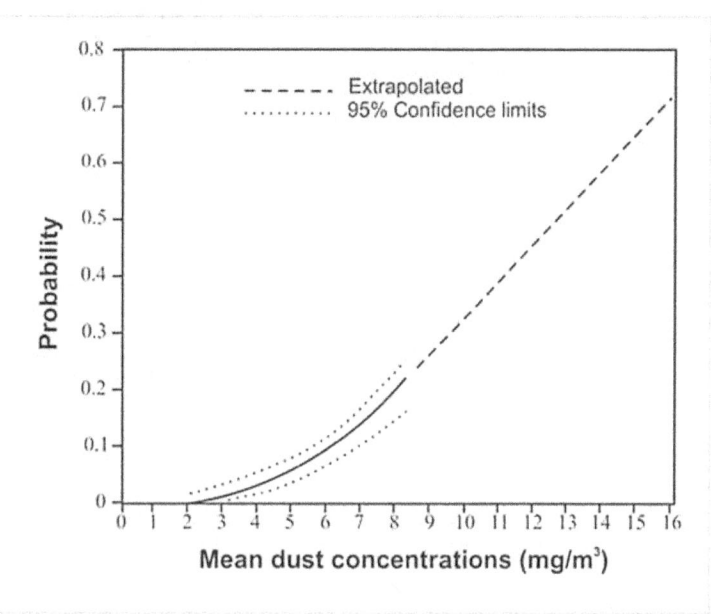

Figure 3. Probability that an individual starting with no pneumoconiosis (category 0/0) will be classified as 2/1 or greater after 35 years of exposure to various concentrations of coal mine dust. [Figure 7–2 of the CCD (1)] (Source: Jacobsen et al. (4)).

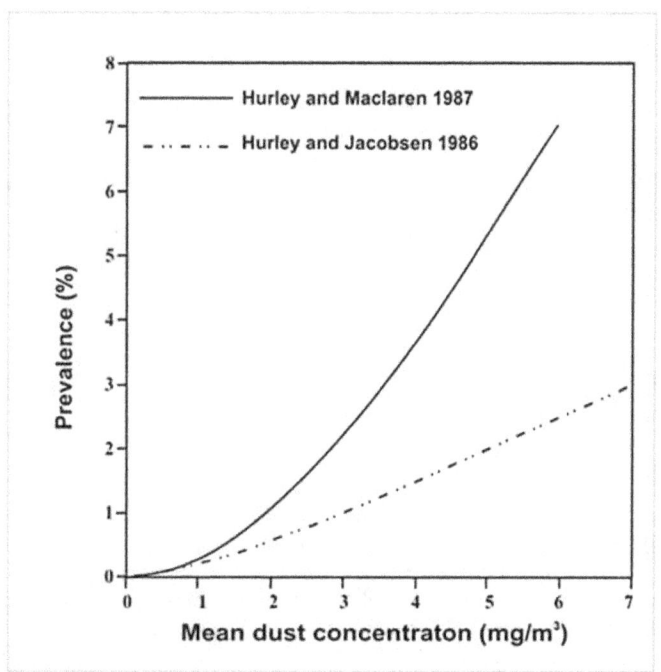

Figure 4. Predicted prevalence of PMF among British coal miners after a 35-year working lifetime by mean concentration of respirable coal mine dust. [Figure 7–6 of the CCD (1)](Source: Hurley and Maclaren (5)).

Table 1. Predicted prevalence of simple CWP and PMF among U.S. or British coal miners at age 58 followingT exposure to respirable coal mine dust over a 40-year working lifetime. (Table 4–6 of the CCD(1))

Study and coal rank	Mean concentration of respirable CMD (mg/m³)	Predicted prevalence (cases/1,000)		
		CWP≥1	CWP≥2	CWP≥3
Attfield and Seixas (6):*				
High-rank bituminous	2.0	253 (204–308)†	89 (60–130)	51 (30–85)
	1.0	116 (88–150)	29 (16–51)	16 (7–36)
Medium/low-rank bituminous	2.0	144 (117–176)	31 (20–49)	14 (7–27)
	1.0	84 (64–110)	17 (9–30)	9 (4–19)
Attfield and Morring (7):‡				
Anthracite	2.0	316 (278–356)	142 (118–172)	89 (69–113)
	1.0	128 (108–152)	46 (35–60)	34 (24–48)
High-rank bituminous (89% carbon)	2.0	282 (250–317)	115 (94–141)	65 (49–85)
	1.0	119 (100–142)	41 (31–54)	29 (20–41)
Medium/low rank bituminous (83% carbon)	2.0	121 (108–136)	40 (33–49)	22 (17–29)
	1.0	74 (62–89)	24 (18–31)	17 (12–24)
Medium/low-rank bituminous (Midwest)	2.0	89 (73–108)	28 (20–39)	15§ (9–26)
	1.0	63 (52–77)	20 (14–27)	14§ (9–21)
Medium/low rank bituminous (West)	2.0	67 (52–86)	15 (8–26)	13§ (7–24)
	1.0	55 (44–68)	14 (10–21)	12§ (8–20)
Hurley and Maclaren				
High-rank bituminous (89% carbon)	2.0	89	29	18
	1.0	40	12	7
Medium/low-rank bituminous (83% carbon)	2.0	65	16	7
	1.0	28	7	3

*Attfield and Seixas (6) define the coal rank groups as follows: 1- high-rank bituminous (89–90% carbon)—central Pennsylvania and southern West Virginia; 2-medium/low-rank bituminous (80–87% carbon)—western Pennsylvania, northern and southwestern West Virginia, eastern Ohio, eastern Kentucky, western Virginia and Alabama; 3- low-rank—western Kentucky, Illinois, Utah, and Colorado.
†Ninety-five percent confidence intervals, where available, are in parentheses under the point estimates for prevalence (cases/ 1000).
‡In Attfield and Morring (7), the predicted prevalences for CWP category 1 or greater and category 2 or greater did not include PMF (correction from CCD original).
§Attfield and Morring (7) define the coal rank groups as follows: 1- anthracite - two mines in eastern Pennsylvania (~93% carbon); 2- medium/low-volatile bituminous (89–90% carbon)—three mines in central Pennsylvania and three in southeastern West Virginia; 3- high-volatile "A" bituminous (80–87% carbon)—16 mines in western Pennsylvania, north and southwestern West Virginia, eastern Ohio, eastern Kentucky and Illinois.

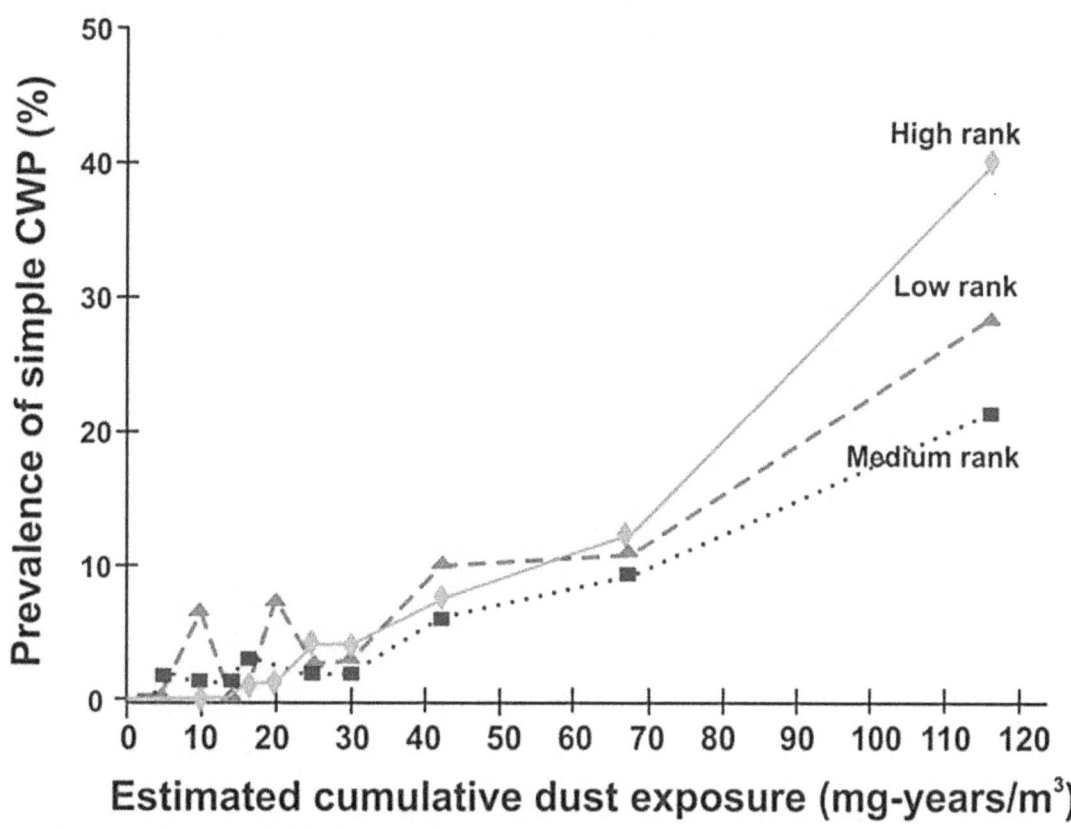

Figure 5. Prevalence of simple CWP category 1 or greater among U.S. coal miners by estimated cumulative coal mine dust exposure and coal rank. [Figure 7–4 of the CCD (1)](Source: Attfield and Seixas (6)).

results provided here are selected to illustrate the main CCD conclusions. For a full understanding of the complete body of knowledge, please see the CCD (1).)

On the basis of the updated body of evidence on the adverse health effects and an evaluation of the technological feasibility of reducing dust exposures, NIOSH recommended that the federal coal mine dust limit be reduced to 1 mg/m³. Critical to this decision were computed excess (exposure-attributable) prevalences of CWP derived from two studies of U.S. coal miners undertaken by NIOSH (CCD Table 7–2 (1); Table 3). Predictions were derived from each study for a working lifetime (i.e., 45 years) exposure at 2 mg/m³ and 1 mg/m³. Another source of information critical to the recommendations was information on predicted excess lung function decrements for a lifetime's exposure to 1 mg/m³ and 2 mg/m³ respectively (CCD Table 7–3 (1); Table 4). Details of the rationale for, and development of, the risk analyses employed in the CCD were subsequently published separately (8). NIOSH also evaluated information from other epidemiologic studies in coming to its recommendations in the CCD. However, because no other studies had quantitative exposure-response information, apart from strengthening the

Table 2. Predicted prevalence of decreased lung function* among U.S. or British coal miners at age 58 following exposure to respirable coal mine dust over a 40-year working lifetime. (Table 4–7 of the CCD (1))

Study and region†	Mean concentration of respirable coal mine dust (mg/m³)	Lung function decrement (% FEV$_1$)	Predicted prevalence (cases/1,000)	
			Never smokers	Smokers
Attfield and Hodous (8)				
East	2.0	<80	141	369
		<65	22	102
	1.0	<80	123	336
		<65	18	87
West	2.0	<80	125	340
		<65	16	80
	1.0	<80	108	309
		<65	13	68
Marine et al. (9):				
	2.0	<80	153	372
		<65	63	173
	1.0	<80	125	314
		<65	52	159

*Decreasing lung function is defined as FEV$_1$ <80% of predicted normal values. Clinically important deficits are FEV$_1$ <80%, which approximately equals the lower limit of normal (LLN), or the fifth percentile (9, 10); and FEV$_1$ <65%, which has been associated with severe exertional dyspnea (11, 12).
†Attfield and Hodous (13) define the following coal rank regions: East—anthracite (eastern Pennsylvania) and bituminous (central Pennsylvania, northern Appalachia [Ohio, northern West Virginia, western Pennsylvania], southern Appalachia [southern West Virginia, eastern Kentucky, western Virginia]), Midwest (Illinois, western Kentucky), and South (Alabama).
‡Conversion from gh/m³ to mg-yr/m³; assumed 1,920 hr/yr for U.S. miners.

case for more stringent regulation, these additional study results did not provide any numerical basis for standard setting. For simplicity, NIOSH recommended one exposure limit for the nation rather than different limits by coal rank, based on technological feasibility of reducing exposures, even though CWP prevalence has been shown to vary according to the rank of the coal in studies of miners in the U.S. and other countries.

In addition to recommending a reduction in the exposure limit for coal mine dust, NIOSH also recommended a change in the exposure limit for crystalline silica dust and the method by which it is enforced. Currently, silica levels are intended to be controlled by a reduction in the level of coal mine dust commensurate with the proportion of the dust that is silica. NIOSH proposed a separate limit for respirable crystalline silica in order to more effectively monitor

Table 3. Excess (exposure-attributable) prevalence of simple CWP or PMF among U.S. coal miners at age 65 following exposure to respirable coal mine dust over a 45-year working lifetime. (Table 7–2 of the CCD (1))

Study and coal rank	Disease category	Cases/1,000 at various mean dust concentrations		
		0.5 mg/m^3	1.0 mg/m^3	2.0 mg/m^3
Attfield and Seixas (6):*				
High-rank bituminous	CWP ≥ 1	48	119	341
	CWP ≥ 2	20	58	230
	PMF	13	36	155
Medium/low-rank bituminous	CWP ≥ 1	27	63	165
	CWP ≥ 2	9	22	65
	PMF	4	10	29
Attfield and Morring (7):†				
Anthracite	CWP ≥ 1	45	120	380
	CWP ≥ 2	17	51	212
	PMF	17	46	167
High-rank bituminous (89% carbon)	CWP ≥ 1	41	108	338
	CWP ≥ 2	15	43	168
	PMF	13	34	114
Medium/low-rank bituminous (83% carbon)	CWP ≥ 1	18	42	111
	CWP ≥ 2	6	15	42
	PMF	4	9	21
Medium/low-rank bituminous (Midwest)	CWP ≥ 1	12	26	64
	CWP ≥ 2	4	9	22
	PMF	1	3	6
Medium/low-rank bituminous (West)	CWP ≥ 1	7	14	32
	CWP ≥ 2	<1	<1	1
	PMF	<1	<1	1

*Attfield and Seixas (6) define the coal rank groups as follows: 1- high-rank bituminous (89–90% carbon)—central Pennsylvania and southern West Virginia; 2- medium/low-rank bituminous (80–87% carbon)—western Pennsylvania, northern and southwestern West Virginia, eastern Ohio, eastern Kentucky, western Virginia and Alabama; 3- low-rank—western Kentucky, Illinois, Utah and Colorado.

†Attfield and Morring (7) define the coal rank groups as follows: 1- anthracite - two mines in eastern Pennsylvania (~93% carbon); 2- medium/low-volatile bituminous (89–90% carbon)—three mines in central Pennsylvania and three in southeastern West Virginia; 3- high-volatile "A" bituminous (80–87% carbon)—16 mines in western Pennsylvania, north and southwestern West Virginia, eastern Ohio, eastern Kentucky and Illinois.

Table 4. Excess (exposure-attributable) prevalence of decreased lung function* among U.S. coal miners at age 65 following exposure to respirable coal mine dust over a 45-year working lifetime. (Table 7–3 of the CCD (1))

Study and region	Lung function decrement	Smoking status	Cases/1,000 at various mean dust concentrations		
			0.5 mg/m^3	1.0 mg/m^3	2.0 mg/m^3
Attfield and Hodus (8):†					
East	<80% FEV$_1$	Never smoked	10	21	44
		Smoker	12	24	51
West	<80% FEV$_1$	Never smoked	9	19	40
		Smoker	11	23	48
East	<65% FEV$_1$	Never smoked	2	5	12
		Smoker	4	8	19
West	<65% FEV$_1$	Never smoked	2	4	9
		Smoker	3	7	15
Seixas et al. (14):	<80% FEV$_1$	Never smoked	60	134	315
		Smoker	68	149	338
	<65% FEV$_1$	Never smoked	18	45	139
		Smoker	27	67	188

*Decreasing lung function is defined as FEV$_1$ <80% of predicted normal values. Clinically important deficits are FEV$_1$ <80%, which approximately equals the lower limit of normal (LLN), or the fifth percentile (9, 10); and FEV$_1$ <65%, which has been associated with severe exertional dyspnea (11, 12).

†Attfield and Hodus (8) define the following coal rank regions: East—anthracite (eastern Pennsylvania) and bituminous (central Pennsylvania, northern Appalachia [Ohio, northern West Virginia, western Pennsylvania], southern Appalachia [southern West Virginia, eastern Kentucky, western Virginia]), Midwest (Illinois, western Kentucky), and South (Alabama).

‡Coal rank was not provided in Seixas et al. (14) However, miners were included from bituminous coal ranks and regions across the United States, as described in Attfield and Seixas (6): 1. High-rank bituminous (89%–90% carbon): central Pennsylvania and southeaster West Virginia; 2. Medium/low-rank bituminous (80%–87% carbon): medium-rank—western Pennsylvania, northern and southwestern West Virginia, eastern Ohio, eastern Kentucky, western Virginia, and Alabama; low-rank—western Kentucky, Illinois, Utah, and Colorado.

and control exposures. The NIOSH recommendations for coal mine dust and crystalline silica dust were explicitly intended for both underground and surface coal operations. In addition, NIOSH recommended enhancing worker medical monitoring, and extending it to surface coal mine workers. An independent advisory committee, which was convened by MSHA in 1996 in response to the NIOSH CCD, affirmed each of the recommendations in the CCD (15).

The following material comprises a summary of results from reports that have been published since 1995 on CWP, other respiratory diseases, cancer outcomes, and overall mortality. In addition, new information summarized from the current NIOSH medical monitoring program for coal miners is included. There are also sections on aspects relating to dust levels, control, and compliance, and on surface coal mining.

2 Coal Workers' Pneumoconiosis

2.1 Surveillance

Over time since 1995 it has become increasingly apparent that the observed prevalence of CWP in U.S. underground coal miners examined in the Coal Miners' X-ray Surveillance Program (CWXSP) was no longer declining as it had from 1969–1995, but had begun increasing. This situation was first noticed in a 2003 CDC/NIOSH report (16). This report also drew attention to the fact that CWP was developing in underground coal miners who had spent all of their working life in a working environment where the dust conditions should have been as mandated by the 1969 Coal Mine Act. Based on findings that showed higher CWP prevalences in certain worker groups, the publication raised concerns about possible excessive dust exposures in certain states, at smaller mines, and by some surface and contract miners.

Reports in 2006 and 2007 called attention to advanced pneumoconiosis in working underground miners in Kentucky (KY) and Virginia (VA) (17, 18); as with the prior report (2003), most of the affected miners had started work after 1969 yet had still developed severe CWP. Possible reasons put forth as explanations for the findings were: 1) inadequacies in the mandated coal mine dust regulations; 2) failure to comply with or adequately enforce those regulations; 3) lack of disease prevention innovations to accommodate changes in mining practices (e.g., thin-seam mining); and 4) missed opportunities by miners to be screened for early disease and take action to reduce dust exposures. Further explanations, noted in other reports included: 5) longer hours being worked by today's coal miners; 6) excessive exposure to crystalline silica, perhaps associated with the mining of thinner seams of coal; and 7) lack of resources for dust control and miner/operator education, particularly in smaller mines (19–21).

To gain a better understanding of the extent of the problem, NIOSH undertook a systematic analysis of rapidly progressive CWP (22). Statistics were derived based on each miner's radiographic steps of progression of CWP using the standard ILO categorical scores standardized to a five-year interval. These data were summarized by county and then plotted to reveal 'hot spots' of rapid disease progression (Figure 6; from Antao et al, 2005 (22)). These tended to be located on the eastern edge of the Appalachian coal field but were particularly concentrated in the southern West Virginia (WV)/western VA/eastern KY tri-state region (central Appalachian region).

In response to these observations, NIOSH undertook a series of field surveys in the hot spot regions in an attempt to enhance the quality of data (i.e., improve participation). The field surveys were undertaken as part of the Coal Workers' X-ray Surveillance Program (CWXSP) administered by NIOSH, as mandated by the 1969 Coal Mine Act. The targeted surveys comprised an "Enhanced Program" to complement the regular CWXSP program. Those findings are included in an overall tabulation that can be found in the NIOSH *2007 Work-related Lung Disease (WoRLD) Surveillance Report* (disseminated in hard copy (23) and

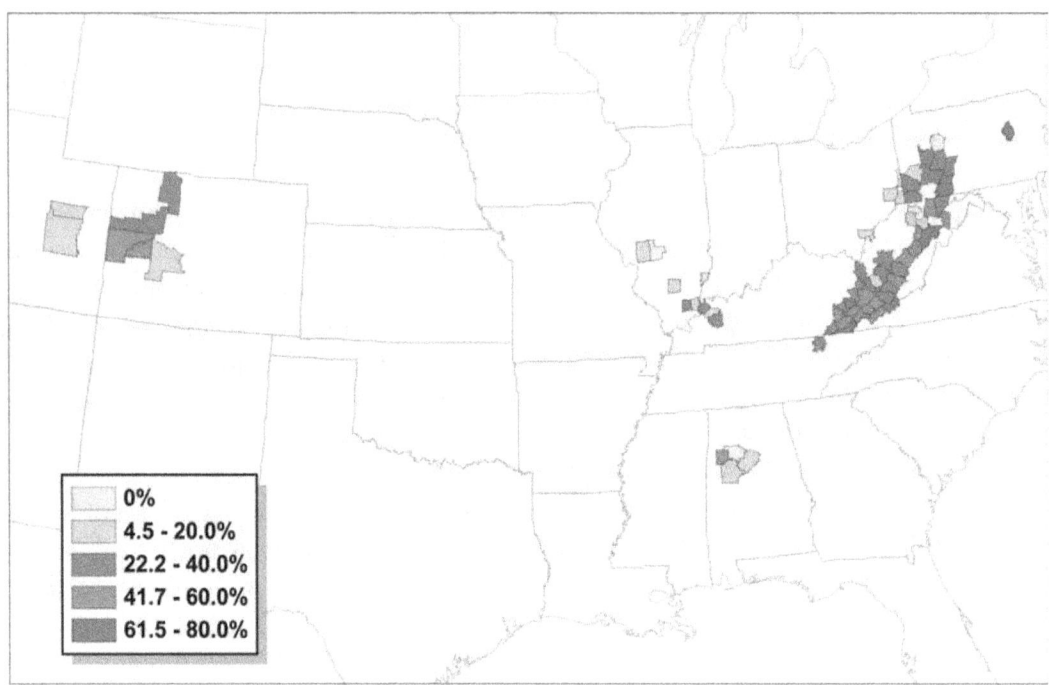

Figure 6. Percentage of evaluated miners with rapidly progressive coal workers' pneumoconiosis by county (not shown are counties with fewer than five miners evaluated). (Source: Antao et al. (22)).

on the NIOSH internet site (24). These results showed that the prevalence of CWP appeared to have stopped declining around 1995–1999, and has risen since then. The trend reversal appears most apparent in the longer-tenured miners. Virtually that entire group had spent their whole working life in dust conditions mandated by the 1969 Coal Mine Act. An updated (unpublished) version of this graphic, taking the data up to 2009, shows the increased CWP prevalence observed over the past decade (Figure 7).

The upward trend visible in Figure 7 for all pneumoconiosis cases (category 1+) is even more evident for PMF (Figure 8). Of particular concern are the prevalence values for the last three five-year periods (1995–2009) for miners with <25 years tenure, which are well above those observed in the early 1990s. In 2005–2009 alone, 69 coal miners examined in the CWXSP were determined to have PMF. Of these, 11 had less than 25 years total tenure in coal mining, and the majority (56, or 81%) were working in the central Appalachian region.

Since the data from 2005 in Figures 7 and 8 were derived, in part, from the special NIOSH surveys targeted at hot spot areas, there could be the concern that the recent CWXSP findings may be upwardly biased, with the implication that the apparent rise in prevalence may be an artifact. However, overall prevalences for 2005–2009 for the Enhanced and regular programs, derived from state-specific prevalences weighted by the participation rates for the whole program from the different states gives rise to figures of 3.2% prevalence for data from the targeted surveys (Enhanced program) compared to 3.1% for data from the regular program. There is therefore no indication whatsoever of any major bias. Moreover, it is

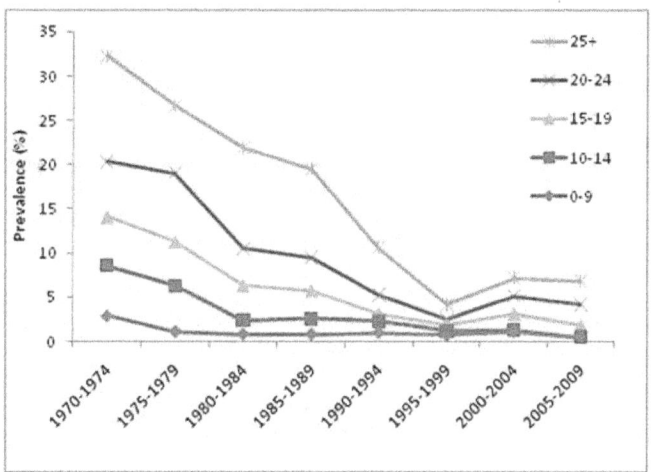

Figure 7. Percentage of miners examined with CWP category 1 or greater from the NIOSH Coal Workers' X-ray Program from 1970–2009, by tenure in coal mining. (Source: NIOSH CWXSP data).

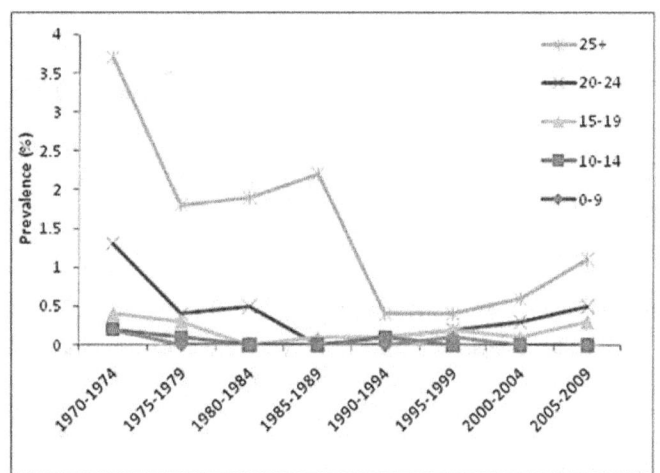

Figure 8. Percentage of miners examined with PMF from the NIOSH Coal Workers' X-ray Program from 1970–2009, by tenure in coal mining. (Source: NIOSH CWXSP data).

clear from Figures 7 and 8 that the upswing in prevalence of CWP was underway before the targeted surveys began in 2005.

The finding of severe CWP in the CWXSP was confirmed among West Virginia coal miners with a report of 138 compensated cases of PMF from 2000–2009 (25). These miners had worked virtually all of their lives under post-1969 Coal Mine Act conditions, and had developed PMF at age 52.6 years on average. Of the 138, 21 had died by publication date.

A number of reports of surveillance information from other countries have emerged since 1995 (26–31). Although mining conditions differ in these other countries, these studies are supportive of the findings in U.S. coal miners.

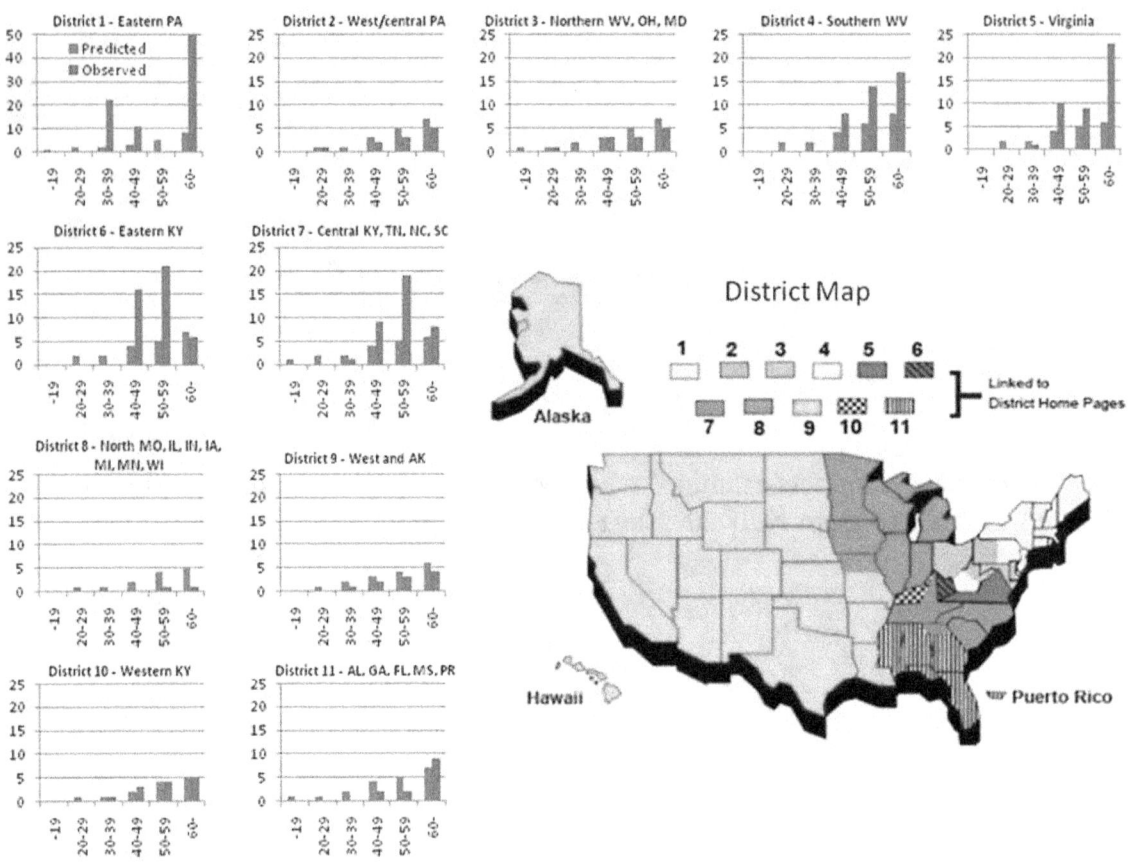

Figure 9. Observed and predicted prevalences (%) of CWP category 1 or greater by age group and MSHA District. (Source: Data from the NIOSH CWXSP for 1995–2009. Charts based on data provided in Suarthana et al (32).)

2.2 Epidemiology

The CWXSP finding of increased prevalence of CWP has prompted a series of analyses by NIOSH aimed at identifying factors that might be causing the increase. In an analysis undertaken as part of this series (32), predicted risk of CWP was derived for each individual who participated in the CWXSP from 2005–2009 using published exposure-response models (7). The models were coal-rank specific, and used both age and cumulative coal mine dust exposure as predictors. The resulting individual risks (lying between 0 and 1) were then summed over subsets of the data and compared with the observed prevalences. (Further models, published later, were also available, but the early relationships were preferred because they were based on greater numbers of observations and had more specific adjustment for coal rank. The later predicted prevalences were somewhat higher than those presented here.) The results tabulated by MSHA region and age are shown here in Figure 9. It is clear that CWP prevalence is less than expected in some regions (observed ≤ predicted) but substantially greater than expected in others (observed > predicted). That is, in the northern Appalachian region and the mid-west and western coal fields the observed prevalences are generally below those predicted in all age groups. However, in the southern WV, eastern and central KY, Tennessee, and VA MSHA regions the observed prevalences are 2–4 times greater than predicted from cumulative coal mine dust exposure and age. Clearly, some factor or factors must be acting differently across the regions to cause this regional pattern.

At least three environmental factors impact the central Appalachian region in this respect. These are: thin seams, small mines, and, for VA, high coal rank. The mining of thin coal seams, which often involves the deliberate cutting and extraction of substantial amounts of (often siliceous) rock overlying or underlying the coal seam, is particularly prevalent in Appalachia (33). NIOSH has been investigating the health implications of possible excessive crystalline silica exposure arising from the cutting of the rock adjacent to the coal seams (34). This analysis used the presence of r-type pneumoconiotic opacities on the chest X-ray as an indicator of crystalline silica exposure. This type of opacity is a radiographic manifestation of nodules in the lung having a typology often associated with excessive exposure to silica dust. An increase in the prevalence of such opacities could then well indicate that miners are more frequently being exposed to crystalline silica dust, or are experiencing exposure to higher levels of silica dust. Increased exposure to crystalline silica dust may well be arising from industry trends, whereby there is greater focus on mining thinner seams of coal as the more productive thicker seams are mined out.

The findings from this study indicated that the proportion of radiographs showing r-type opacities increased during the 1990s, and particularly after 1999, in KY, VA, and WV, compared to the 1980s. They could potentially be explained by an increase in the frequency and/or intensity of silica exposure among underground coal miners. This hypothesis was confirmed by evidence from dust sampling in mines in that region indicating that excessive silica exposures are occurring (35). In the CCD, NIOSH not only recommended that compliance procedures for crystalline silica be made more effective, but that the exposure limit be reduced. In the light of these epidemiological findings (34), therefore, this recommendation remains appropriate and even more urgent. A report on British coal miners also associated an increase in CWP prevalence with rock cutting (36).

In another NIOSH analysis (19) trends in CWP prevalence were examined by mine size (i.e., employment). The hypothesis investigated was that smaller mines lack resources in many areas for the full protection of workers, including dust suppression and up-to-date knowledge of means to prevent disease development. (It may also be that smaller mines tend to be those mining the thinner coal seams.) The results show that CWP prevalence is increasing in mines of all sizes, but the trend is much more obvious and much greater among miners employed in smaller coal mines.

Of the other factors listed above that may be contributing to the rise in CWP prevalence, increased working hours gives rise to special concern. Overall, U.S. coal miners are working longer hours. Figure 10, derived from data collected by MSHA shows a steady increase in the number of hours worked. This increase appears to arise from not only working longer shifts (e.g., 10 or 12 hours), but from working on weekends as well. Although no epidemiologic data exist that implicate longer hours as

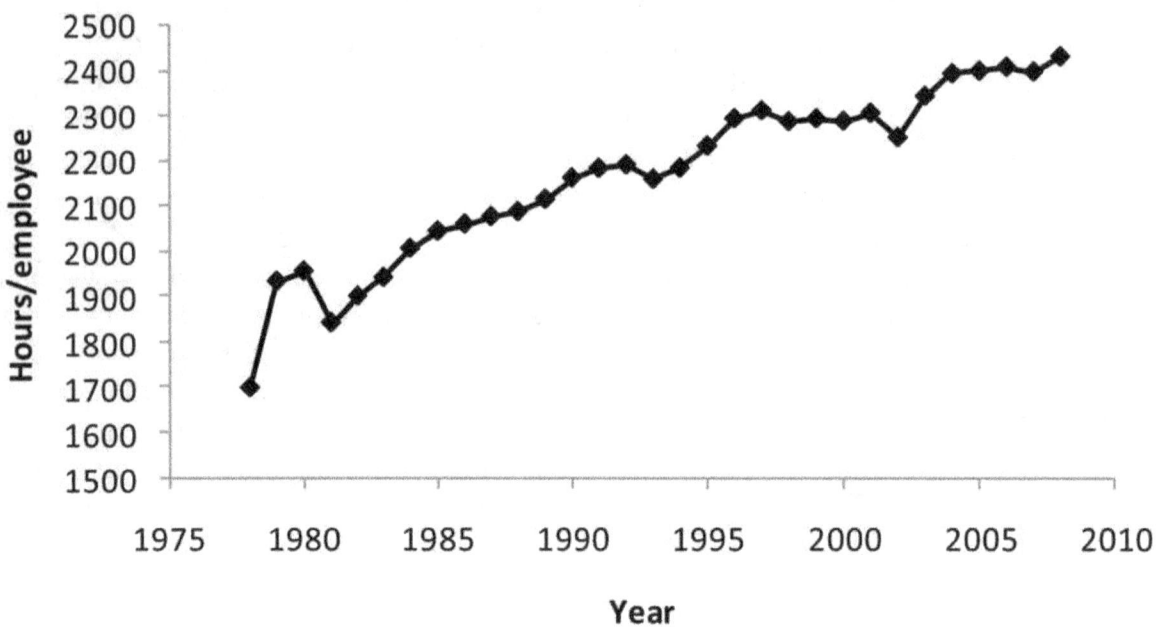

Figure 10. Hours worked per underground coal miner, 1978–2008. (Source: http://www.msha.gov/ACCINJ/BOTHCL.HTM).

a contributory causative factor for CWP, working longer hours leads to the inhalation of more dust into the lungs. For example, working 12 hours leads to 50% more dust entering the lung compared to a regular 8-hour shift, assuming all other factors are equal (e.g., exposure concentration and breathing rates). Additionally, working longer workshifts reduces the time available between workshifts for the process of clearing the dust deposited in the lungs. Unfortunately, the available information on working hours in U.S. coal miners is not miner-specific but rather by coal mine, substantially reducing the validity of a formal analysis of this hypothesis. A report on British miners concluded that longer working hours were a factor in causing an increase in CWP prevalence at two mines (36). In the CCD, NIOSH recommended reducing dust exposures below the 1 mg/m^3 REL for work shifts exceeding 40 hr/week (using the method of Brief and Scala (37)). This approach has also been recommended for British coal mines (38).

Finally, productivity per hour worked also increased from 1978–2000, although it has since declined (Figure 11). Of course, these increases in productivity (and, presumably, increased potential for dust generation) should have been met by commensurate increases in dust prevention measures (e.g., ventilation and water sprays) in order to maintain compliance with the permissible exposure limit. Superficially, the current data appear to confirm this, in that airborne dust levels have apparently not risen during that period (Figure 12). However, the veracity of coal mine dust data has been challenged in the past (39). Moreover, the discovery of abnormal white centers in the dust sampling filters prompted a special inspection

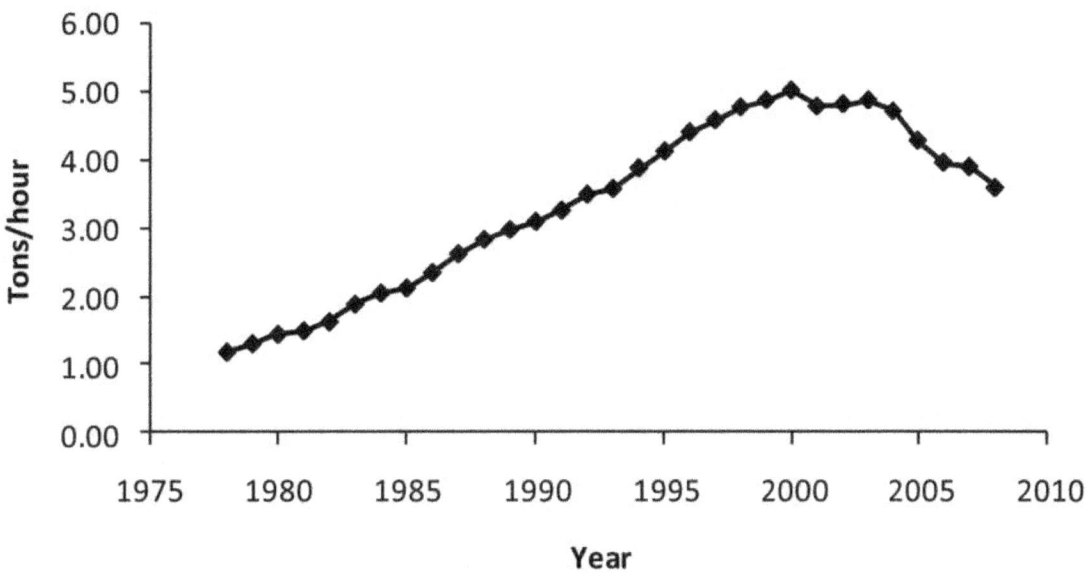

Figure 11. Tons produced per hour worked at underground coal mines. 1978–2008. (Source: http://www.msha.gov/ACCINJ/BOTHCL.HTM).

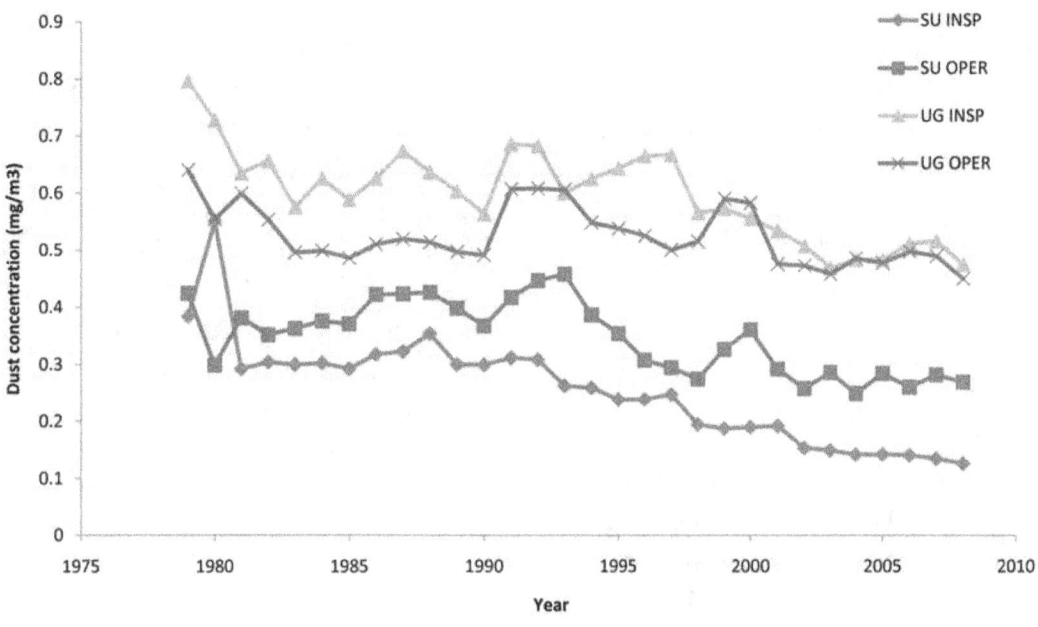

Figure 12. Respirable coal mine dust: Geometric mean exposures by type of mine (UG=underground, SU=surface), MSHA inspector (INSP) and mine operator (OPER) samples. [MSHA coal mine inspector and mine operator dust data].

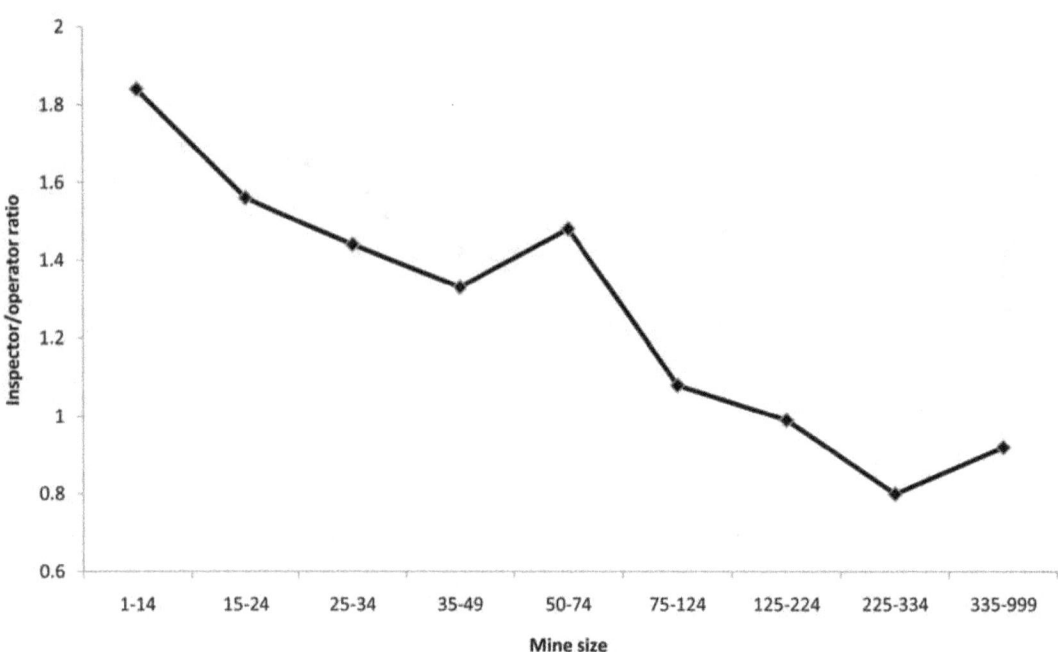

Figure 13. Ratio of special inspection sample values to preceding operator compliance sample values by mine size. (Source: MSHA Report of the Statistical Task Team of the Coal Mine Respirable Dust Task Group. (40)).

program (40) that showed that dust levels from operator samples consistently were lower than those from MSHA inspector samples, and that these differences were greater the smaller the mine (Figure 13). As with hours worked, there is a lack of reliable productivity data linkable with the health outcome data in order to investigate this issue further.

2.3 Mortality

A report on temporal patterns in pneumoconiosis mortality in the U.S. showed a substantial decline in numbers of deaths from CWP between 1968 and 2000 (41). This decline is consistent with the reductions in dust level mandated by the 1969 Coal Mine Act. A major additional factor contributing to the declining number of CWP deaths is the diminishing coal mining workforce in the U.S. Figure 14 (Figure 1, from CDC (42)) shows the CWP death rate results extended to 2006. A similar situation has been observed in other developed countries, e.g., Australia (26). Recent U.S. results have shown, however, a disconcerting increase in years of potential life lost (YPLL) due to CWP in the U.S. since 2002 (42). Not only has the YPLL been increasing in younger CWP decedents (<65 years old), but the YPLL per CWP decedent has also been increasing over those same years (Figure 15; Figure 2 from CDC (42)). This may be related to the observed increase in CWP prevalence observed in recent years as noted earlier.

The post-1995 period saw the publication of a number of mortality analyses that augmented the earlier mortality findings on coal miners.

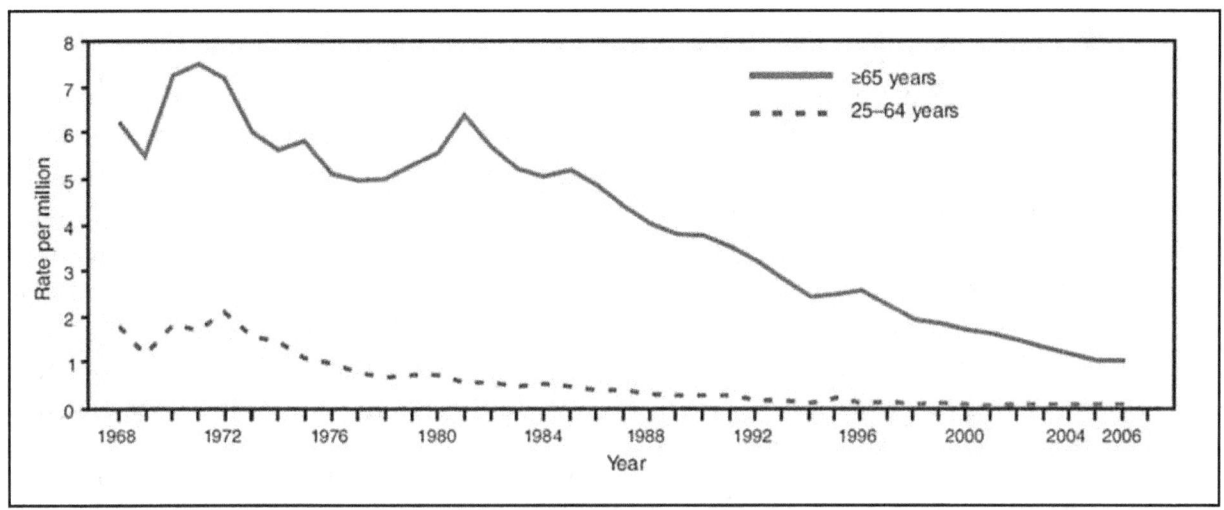

Figure 14. Age-adjusted death rates (per million) for decedents age ≥25 years with coal workers' pneumoconiosis as the underlying cause of death—United States, 1968–2006. (Source: CDC (42)).

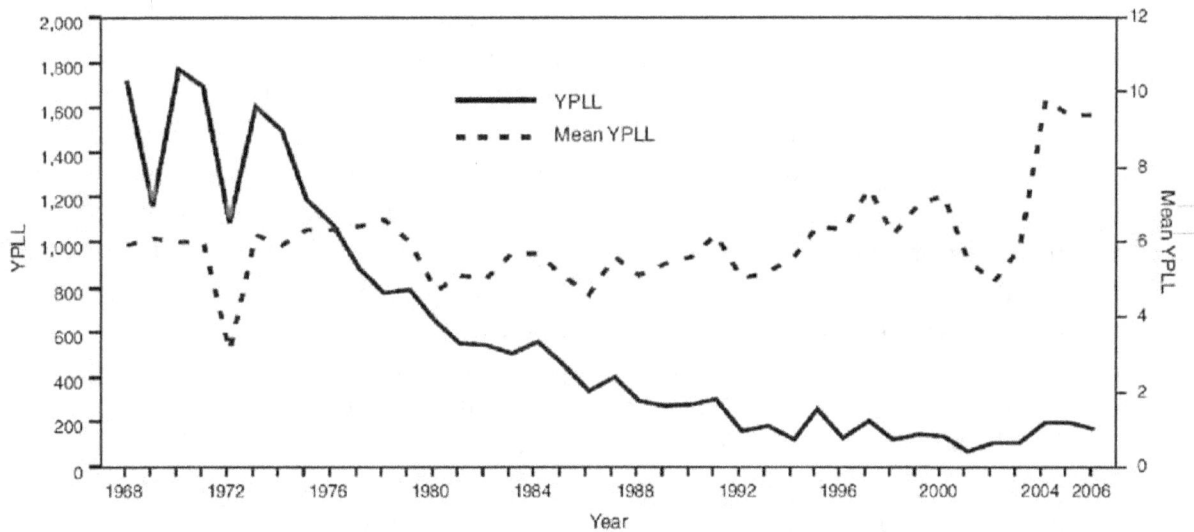

Figure 15. Years of potential life lost (YPLL) before age 65 and mean YPLL per decedent for decedents aged ≥25 years with coal workers' pneumoconiosis as the underlying cause of death—United States, 1968–2006. (Source: CDC (42)).

Most of those studies outside of the United States and United Kingdom did not have quantitative measurements of dust exposure. However, they do support previous findings concerning the overall increased mortality of coal miners and the additional risk imposed by the development of CWP (43–45). Studies using quantitative exposures showed that mortality from CWP increased with increasing cumulative exposure to coal mine dust (46). The British study (47) included exposure estimates for respirable quartz; cumulative exposure to respirable coal mine dust and respirable quartz were each highly significant predictors of pneumoconiosis mortality, although the relationship was stronger with coal dust than quartz. Respirable quartz exposure was associated with a small but statistically significant relative risk for lung cancer mortality (47).

2.4 Toxicology

Although coal mine dust and crystalline silica dust remain the two exposures of primary concern for environmental control, the post-1995 period has seen the publication of results from analyses aimed at eliciting information on what constituents of coal mine dust predict CWP development. These include: 1) free radicals, in which particles from freshly-fractured siliceous rock have been found to be more fibrogenic than aged particles (48); 2) particle occlusion, in which clay present in the rock strata can surround the silica particles and render them less toxic (49); and 3) bioavailable iron, which has been found to predict coal mine dust toxicity (50, 51).

McCunney et al. (52), favored the third explanation (bioavailable iron) and downplayed the role of quartz in the etiology of coal workers' pneumoconiosis. However, in an analysis of lung inflammatory cell counts from bronchoalveolar lavage in coal miners and non-miners, Kuempel et al. (53) showed that quartz dust (as either cumulative exposure or estimated lung burden) was a significant predictor of pulmonary inflammation and radiographic category of simple CWP. Cumulative coal dust exposure did not significantly add to those predictions, which may have been due to the high correlation between the coal and quartz cumulative exposures, such that separate effects for these two dusts could not be clearly demonstrated.

Against this, epidemiologic research has not demonstrated a strong effect of crystalline silica on CWP development in situations where silica levels are low. Rather, the level of coal mine dust, per se, has been the strongest predictor of CWP. However, the work of Laney et al. (34), as noted above, showed clear evidence of an increase in r-type radiographic opacities (typically associated with silicosis) and rapid progression of pneumoconiosis among U.S. coal miners in Kentucky, Virginia, and West Virginia, suggesting that they were exposed to excessive levels of respirable crystalline silica, and were thus at risk of silicosis. As noted previously, dust sampling results support this hypothesis (23). There is, therefore, the clear need to minimize exposure to silica dust, especially for those jobs involving drilling or cutting sandstone and other siliceous rock. Moreover, as noted above, this is particularly pertinent because changing mining conditions might be leading to an increase in the potential for exposure to silica dust.

Page and Organiscak (54) linked the issue of coal rank, a known risk factor for CWP development in the U.S., Britain, and Germany, with the potential for higher levels of free radicals to be encountered where such coals are mined, and noted above by Dalal (48) and others to have greater levels of cytotoxicity.

2.5 Risk Analysis

Kuempel et al. (13) describe in more detail the risk analyses provided in the NIOSH CCD, including the excess (exposure-attributable) prevalence of CWP and PMF in underground coal miners exposed to various levels of coal mine dust for a working lifetime (as shown in the CCD and also presented here in Table 1). More recent risk estimates have been provided from research on British coal miners (Figure 16, from Figure 1 of Soutar et al. (55)). The latter apply to coal composed of 86.2% carbon (coal rank) and to underground coal miners who work 40 years at the designated coal mine dust level. Risks of PMF range from 0.8% at 1.5 mg/m³ to about 5% at 6 mg/m³, while risks of category 2 or greater CWP range from about 1.5% at 1.5 mg/m³ to about 9% at 6 mg/m³. Note that due to the different ways in which the risk estimates are derived, these are not directly comparable with those from U.S. studies shown in Table 1. However, the findings are consistent with those from U.S. studies in indicating that even at the lower coal mine dust levels recommended by NIOSH, and as noted in the CCD, some incidence of CWP would still be expected, especially among miners of higher rank coal.

Soutar et al. (55) also provide information on the risk of silicosis in underground coal miners. Their findings were developed from observations at one mine in which unusually high concentrations of crystalline silica dust occurred periodically (56). In their analysis, the authors chose to divide the analysis between exposures < 2 mg/m³ and ≥ 2 mg/m³. This dichotomy, in the authors' presentation, was associated with more rapid development of silicosis in the ≥ 2 mg/m³ exposure range compared to chronic silicosis development at exposures < 2 mg/m³. The findings indicate that short excursions to high silica dust intensities are considerably more hazardous than the same level of cumulative exposure at a lower intensity. They therefore demonstrate that mining situations involving the cutting of rock should be avoided if at all possible, or if necessary, that all precautions should be taken to minimize dust exposures. The findings for <2 mg/m³ (which apply to most coal mining environments that do not involve direct rock cutting) are given in Figure 17 (Figure 3 of Soutar et al. (55)).

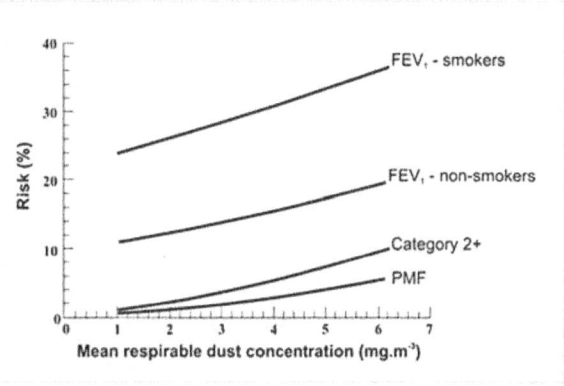

Figure 16. Risks at age 58–60 after 35–40 working years of: PMF; category 2 or greater (2+); 993 ml deficit of FEV_1 in nonsmokers; 993 ml deficit of FEV_1 in smokers. (Source: Soutar et al. (55)).

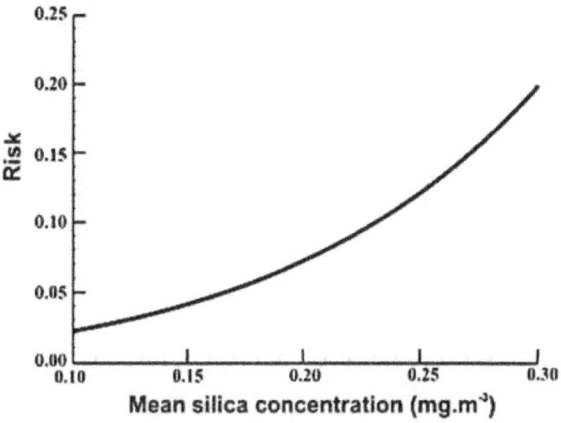

Figure 17. Risks for category 2 silicosis in relation to respirable silica concentration (<2 mg/m³) averaged over 15 years. (Source: Soutar et al. (55)).

3 Other Respiratory Disease Outcomes

Coggon and Taylor (57), in an extensive review, concluded that the "...balance of evidence points overwhelmingly to an impairment of lung function from exposure to coal mine dust, and this is consistent with the increased mortality from COPD that has been observed in coal miners." Findings on COPD and related outcomes in coal miners since 1995 (58–64) have continued to support their conclusion (which was largely based on pre-1995 information). The findings have also identified other risk factors in coal mining for pulmonary disease development. These include work in roof bolting, exposure to explosive blasting fumes, and exposure to dust control spray water previously stored in holding tanks (65).

The post-1995 findings have also elucidated patterns of lung function decline in coal miners, indicating that new miners tend to suffer more severe declines on starting work, after which the declines attenuate somewhat. This finding, derived initially from the analysis by Seixas et al. (66), was explored further by Henneberger and Attfield (58), who confirmed that the temporal pattern of lung function decline was different in newly-hired coal miners as compared to experienced miners. A possible reason for this could be a healthy worker survival effect. A study to explore this issue further, undertaken on new Chinese coal miners, confirmed that starting work in coal mining led to large initial drops in lung function, after which lung function declined at a lesser rate (67). In a follow-up analysis, the researchers reported that the development of respiratory symptoms consistent with bronchitis contributed to the early declines in lung function (68).

A recently published mortality study from the United States (46) comprised a longer follow-up of a study on the same cohort of underground coal miners published in 1995 (69). It showed that mortality from chronic airway obstruction (CAO) was elevated. Smoking, pneumoconiosis, coal rank region, and cumulative coal mine dust exposure were all predictors of mortality from CAO. Dust exposure effects were observed within the never-smoker subset of the cohort. The observed dust-related relative risks for CAO were similar to those for pneumoconiosis. The findings showed dust-related effects for chronic bronchitis and emphysema as well as CAO. A recent British study re-affirmed that mortality from COPD was related to coal mine dust exposure (47). Finally, the implications of COPD (due to coal mine dust exposure as well as smoking) in causing increased mortality was explored by examining mortality risk in relation to rates of ventilatory function decline in coal miners (70). Rates of ventilatory decline 2–3 times the normal age-related decline were associated with distinct increases in subsequent mortality.

Past pathologic studies have shown that emphysema severity in coal miners is related to dust exposure. Recent studies on South African and U.S. coal miners confirmed these findings (71, 72). Important additional information on this topic, using quantitative estimates of both coal mine dust exposure and smoking amount, has been recently published by Kuempel et al. (73). These authors found a highly significant relationship between cumulative exposure to respirable coal mine dust and emphysema severity at autopsy, controlling for effects of smoking, age, and other variables. The effect of dust exposure was similar in magnitude to that of smoking, and was seen in the never-smoking subgroup. In a further analysis, Kuempel et al. established that exposure to coal mine dust can produce clinically important levels of emphysema in coal miners (74).

The above findings support the CCD's recommendation to reduce the permissible coal mine dust exposure limit in underground coal mines to prevent the development of COPD, the associated severe declines in lung function, and the ensuing premature mortality.

There have been several reports of interstitial disease associated with exposure to coal mine dust, perhaps representing a manifestation of CWP, although little systematic research on this topic has been undertaken (75, 76).

3.1 Risk Analysis

Kuempel et al. (13) describe in more detail the risk analyses summarized in the NIOSH CCD, including the excess (exposure-attributable) prevalence of lung function deficits in underground coal miners exposed to various levels of coal mine dust for a working lifetime (CCD Table 4–7 (1); Table 2). More recent risk estimates from research on U.K. coal miners have been published (Figure 15; Figure 1 of Soutar et al. (55)). They apply to coal composed of 86.2% carbon (coal rank) and to underground coal miners who work 35 years at specified coal mine dust levels ranging from 1 to 6 mg/m^3. Risks of a deficit of approximately 1 liter in forced expiratory volume in 1 second (FEV_1) among never smokers range from 10% at zero dust exposure to about 19% at 6 mg/m^3. The concomitant risks for smokers range from about 22% to 36%, respectively. Note that due to the different ways in which the risk estimates have been derived, these are not directly comparable with those shown from U.S. studies shown in Table 2. However, they are consistent with findings from U.S. studies in that even at the 1 mg/m^3 coal mine dust exposure limit recommended by the CCD, some occupational effect on ventilatory function is expected.

4 Cancer Outcomes

Two cancer outcomes—lung cancer and stomach cancer—have been of particular interest with respect to work in coal mining. Lung cancer has been suspected to arise in coal miners because of their exposure to crystalline silica dust, which has been determined to be a Group I carcinogen by the International Agency for Research on Cancer, at least in some occupational settings (77). However, findings in coal miners have been conflicting and have not strongly supported a relationship between coal mine dust exposure and lung cancer. The post-1995 findings continue this picture. No overall excess or relationship with increasing dust exposure was seen in lung cancer mortality in a study of U.S. underground coal miners (46). However, this study, lacking silica dust exposure measurements, could not effectively evaluate the hypothesis of interest. In contrast, a recent British study that did include cumulative crystalline silica dust exposures found a weak relationship of silica exposure with lung cancer mortality (47). A recent development in this regard is the finding that lung-deposited silica or coal dust inhibits the induction of cytochrome P4501A1 by polycyclic aromatic hydrocarbons (PAH) (78–80). It is hypothesized that the resulting lower cytochrome activity might to some extent counteract the carcinogenic effects of tobacco smoke by limiting metabolism of PAH in tobacco smoke into carcinogenic metabolites. This may explain the lack of clear findings on dust exposure and lung cancer in coal mining.

There have been occasional reports of elevated stomach cancer mortality among coal miners. The post-1995 results from various reports have not confirmed these findings. In particular, no relationship was detected in the two studies having quantitative exposure measurements (46, 47).

5 Dust Exposure Levels, Control, and Compliance

5.1 Dust Exposure Levels

Overall trends in reported coal mine dust and crystalline silica exposure levels for the United States are shown in Figures 12 and 18. These follow the format of the *2007 WoRLD Surveillance Report* (23) Figures 2–6 and 3–5a, but are updated to 2008. The data in both figures imply that dust levels have declined over time, with those from recent years being about 75% of those around 1980, overall. This has occurred over a time period when underground production levels from both longwall and continuous miner operations significantly increased. However, the reductions vary depending on the type of mine, the source of the data, and the type of dust. The biggest reduction in reported levels was for coal mine dust at surface mines as sampled by inspectors (recent levels are ~40% of those around 1980). The smallest decline was for silica levels in underground mines, where there has been essentially no change over the time period (recent levels are ~98% of those in the early 1980s). Overall, levels of both coal mine dust and crystalline silica dust were reported to be higher in underground mines than in surface mines.

5.2 Dust Exposure Assessment

The primary advance since 1995 in the dust assessment arena has been the development of a

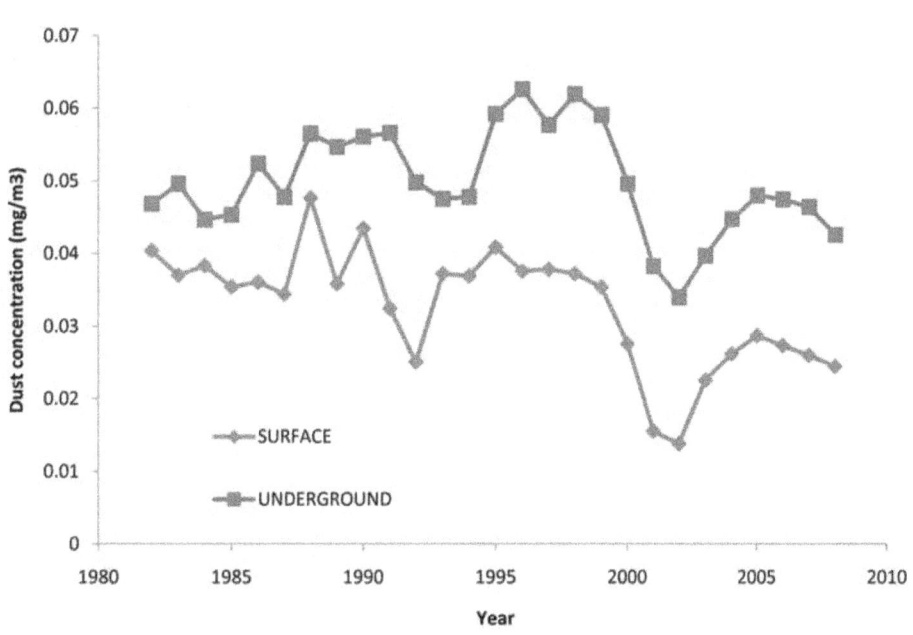

Figure 18. Respirable quartz dust: Geometric mean exposures by type of coal mine (operator and inspector data combined). [MSHA coal mine inspector and mine operator dust data].

continuously-measuring personal dust monitor (PDM) (81). The PDM enables within-shift assessment of dust exposures, permitting prompt action to intervene and reduce excessive levels. Conventional practices that rely on the gravimetric assessment of dust collected on air sample filters preclude speedy remediation because the delay in obtaining results from the dust laboratory could mean that miners continue to be over-exposed before any indication of a problem is available from the laboratory results. The personal dust monitor is now a commercially available product and, as its use is adopted by mines, more timely and targeted interventions to reduce dust exposures will be possible. In 2010, the Mine Safety and Health Administration published new rules that provide for the approval and use of the PDM, in addition to the Coal Mine Dust Personal Sampler Unit, for determining the concentration of respirable dust in coal mine atmospheres (82).

5.3 Compliance Policy and Procedures

The federal policies and procedures for regulating underground coal mine dust levels have been the subject of criticism from their introduction in 1969. Since 1995 further critiques have been published (83, 84). The first publication provided a historical review, the basic argument being that the problems were intrinsic to a process in which an industry essentially regulates itself (i.e., through performing the airborne sampling upon which citations are based). The second publication revisited an issue that was addressed in the NIOSH CCD, in which it was recommended that MSHA not apply any upward adjustment of the REL for instrument uncertainty. Information indicating that reported dust levels from mine operator sampling were systematically lower than those obtained by mine inspectors during unannounced visits to mines to measure exposures has been published by MSHA (40).

6 Surface Coal Mining

Studies published prior to the NIOSH CCD showed that U.S. surface coal miners (particularly workers on drill crews) were at risk of developing CWP (or silicosis). There was also evidence that ventilatory function was reduced in relation to the number of years worked as drill operators or helpers. Since the CCD, a British study has reported evidence of CWP among workers in the dustier jobs and an association with intensity of exposure (85). Dust exposures were generally <1 mg/m^3. In the United States, a relationship between tenure in surface coal mining jobs and prevalence of CWP (ILO category 1/0 or greater, and PMF) was reported (16).

7 Summary

A considerable body of literature has been produced from studies of coal miners and the coal mining environment since 1995, both in the United States and elsewhere. Many of the newer publications, particularly those from other countries, lack quantitative dust exposure measurements, prohibiting full and valid examination of exposure-response relationships. Nevertheless, their findings all support early findings on British and U.S. coal miners, reinforcing the generally-accepted understanding that exposure to coal mine dust can give rise to various respiratory diseases, and that those diseases can cause disability and premature mortality. The remainder of the newer publications that do have quantitative exposure data report findings that refine or augment the fundamental exposure-response results summarized in the CCD.

Overall, the following conclusions can be made:

1. No new findings have emerged since 1995 that contradict the basic summarization of the respiratory health effects of coal mine dust and their relationship with dust exposures described in the CCD (1).

2. No new findings have emerged that substantially modify the basic understanding of coal mine dust exposure and its impact on respiratory health described in the CCD.

3. The new findings that have emerged strengthen prior results and also refine or add further knowledge on disease patterns and etiology described in the CCD.

4. Overall, the logical basis for recommendations concerning prevention of occupational respiratory disease among coal miners remains essentially unaffected by the newer findings that have emerged since publication of the CCD.

New findings of particular note are:

1. After a long period of declining CWP prevalence, recent federal surveillance data indicate that the prevalence is rising.

2. Coal miners are developing severe CWP at relatively young ages.

3. There is some indication that the mortality of younger coal miners from CWP is increasing. These workers would have been employed all of their working lives in environmental conditions mandated by the 1969 Coal Mine Act.

4. The pattern of CWP occurrence across the nation is not uniform; hot spots of disease appear to be concentrated in the central Appalachian region of southern WV, eastern KY, and western VA.

5. The cause of this resurgence in disease is likely multifactorial. Possible explanations include excessive exposure due to increases in coal mine dust levels and duration of exposure (longer working hours), and increases in crystalline silica exposure (see below). As indicated by data on disease prevalence and severity, workers in smaller mines may be at special risk.

6. Given that the more productive seams of coal are being mined out, a transition by the industry to mining thinner coal seams and those with more rock

intrusions is taking place and will likely accelerate in the future. Concomitant with this is the likelihood of increased potential for exposure to crystalline silica, and associated increased risk of silicosis, in coal mining.

In summary, every effort needs to be made to reduce exposures both to respirable coal mine dust and to respirable crystalline silica. As recommended in the CCD, the latter task requires establishing a separate compliance standard in order to effectively limit exposure to silica dust.

References

1. National Institute for Occupational Safety and Health. [1995]. Criteria for a recommended standard: Occupational exposure to coal mine dust. DHHS (NIOSH) Publication No. 95–106. Washington, DC, National Institute for Occupational Safety and Health.

2. Attfield MD, Wagner G [1992]. Respiratory disease in coal miners. In: Rom WN ed. Environmental and occupational medicine. Boston, MA: Little, Brown and Company, pp. 325–344.

3. International Labour Office [1980]. International classification of radiographs of pneumoconiosis 1980 edition. Occupational Safety and Health Series no. 22 (Rev. 80). Geneva: International Labour Office, pp. 1–48.

4. Jacobsen M, Rae S, Walton WH, Rogan JM [1971]. The relation between pneumoconiosis and dust exposure in British coal mines. In: Walton WH ed. Inhaled particles III. Old Woking, England: Unwin Brothers, pp. 903–919.

5. Hurley JF, Maclaren WM. [1987]. Dust-related risks of radiological changes in coalminers over a 40-year working life: Report on work commissioned by NIOSH. TM/79/09. Edinburgh, Scotland, Institute of Occupational Medicine.

6. Attfield MD, Seixas NS [1995]. Prevalence of pneumoconiosis and its relationship to dust exposure in a cohort of U.S. bituminous coal miners and ex-miners. Am J Ind Med 27:137–151.

7. Attfield MD, Morring K [1992]. An investigation into the relationship between coal workers' pneumoconiosis and dust exposure in U.S. coal miners. Am Ind Hyg Assoc J 53:486–492.

8. Attfield MD, Hodous TK [1992]. Pulmonary function of U.S. coal miners related to dust exposure estimates. Am Rev Respir Dis 14:605–609.

9. Marine WM, Gurr D, Jacobsen M [1988]. Clinically important respiratory effects of dust exposure and smoking in British coal miners. Am Rev Respir Dis 137:106–112.

10. Soutar C, Campbell S, Gurr D, Lloyd M, Love R, Cowie H, Cowie A, Seaton A [1993]. Important deficits of lung function in three modern colliery populations—relations with dust exposure. Am Rev Respir Dis 147:797–803.

11. Boehlecke B [1986]. Laboratory assessment of respiratory impairment for disability evaluation. DHHS (NIOSH) Publication No. 86–102. In: Merchant JA ed. Cincinnati, OH: U.S. Department of Health and Human Services.

12. ATS [1991]. Lung function testing: selection of reference values and interpretive strategies. Am Rev Respir Dis 144:1202–1218.

13. Kuempel ED, Smith RJ, Attfield MD, Stayner LT [1997]. Risks of occupational respiratory diseases among U.S. coal miners. Appl Occup Environ Hyg 12:823–831.

14. Seixas NS, Robins TG, Attfield MD, Moulton LH [1993]. Longitudinal and cross sectional analyses of exposure to coal mine

dust and pulmonary function in new miners. Br J Ind Med 50:929–937.

15. U.S. Department of Labor [1996]. Report of the Secretary of Labor's Advisory Committee on the Elimination of Pneumoconiosis among Coal Mine Workers. Washington, DC: US Department of Labor,

16. CDC [2003]. Pneumoconiosis prevalence among working coal miners examined in federal chest radiograph surveillance programs—United States, 1996-2002. MMWR Morb Mortal Wkly Rep 52:336–340.

17. CDC [2006]. Advanced cases of coal workers' pneumoconiosis—two counties, Virginia, 2006. MMWR Morb Mortal Wkly Rep 55:909–913.

18. CDC [2007]. Advanced pneumoconiosis among working underground coal miners—Eastern Kentucky and Southwestern Virginia, 2006. MMWR Morb Mortal Wkly Rep 56:652–655.

19. Laney AS, Attfield MD [2010]. Coal workers' pneumoconiosis and progressive massive fibrosis are increasingly more prevalent among workers in small under ground coal mines in the United States. Occup Environ Med 67:428–431.

20. Loomis D [2010]. Basic protections are still lacking. Occup Environ Med 67:361.

21. Seaton A [2010]. Coal workers' pneumoconiosis in small mines in the United States. Occup Environ Med 67:364.

22. Antao VC, Petsonk EL, Sokolow LZ, Wolfe AL, Pinheiro GA, Hale JM, Attfield MD [2005]. Rapidly progressive coal workers' pneumoconiosis in the United States: geographic clustering and other factors. Occup Environ Med 62:670–674.

23. National Institute for Occupational Safety and Health [2008]. Work-related lung disease surveillance report 2007, Volume 1. DHHS (NIOSH) Publication No. 2008–143a. Cincinnati, OH, National Institute for Occupational Safety and Health.

24. National Institute for Occupational Safety and Health [2010]. Work-related lung disease (WORLD) surveillance system. http://www2a.cdc.gov/drds/WorldReportData/

25. Wade WA, Petsonk EL, Young B, Mogri I [2011]. Severe Occupational Pneumoconiosis Among West Virginia Coal Miners: 138 Cases of Progressive Massive Fibrosis Compensated Between 2000–2009. Chest DOI 10.1378/chest. 10-1326:1–14.

26. Smith DR, Leggat PA [2006]. 24 years of pneumoconiosis mortality surveillance in Australia. J Occup Health 48:309–313.

27. Baur X, Latza U [2005]. Non-malignant occupational respiratory diseases in Germany in comparison with those of other countries. Int Arch Occup Environ Health 78:593–602.

28. Naidoo RN, Robins TG, Solomon A, White N, Franzblau A [2004]. Radiographic outcomes among South African coal miners. Int Arch Occup Environ Health 77:471–481.

29. Marek K, Lebecki K [1999]. Occurrence and prevention of coal miners' pneumoconiosis in Poland. Am J Ind Med 36: 610–617.

30. Nguyen AL, Matsuda S [1998]. Pneumoconiosis problem among the Vietnamese coal mine workers. J UOEH 20:353–360.

31. Parihar YS, Patnaik JP, Nema BK, Sahoo GB, Misra IB, Adhikary S [1997]. Coal workers' pneumoconiosis: a study of prevalence in coal mines of eastern Madhya

Pradesh and Orissa states of India. Ind Health *35*:467–473.

32. Suarthana E, Laney AS, Storey E, Hale JM, Attfield MD. Coal Workers' Pneumoconiosis in the United States: Regional Differences 40 Years after Implementation of the 1969 Federal Coal Mine Health and Safety Act. Occup.Environ. Med 2011 (In Press)

33. Peters RH, Fotta B, Mallett LG [2001]. The influence of seam height on lost-time injury and fatality rates at small underground bituminous coal mines. Appl Occup Environ Hyg *16*:1028–1034.

34. Laney AS, Petsonk EL, Attfield MD [2009]. Pneumoconiosis among under ground bituminous coal miners in the United States: is silicosis becoming more frequent? Occup Environ Med

35. Pollock DE, Potts JO, Joy GJ [2010]. Investigation into dust exposures and mining practices in mines in the southern Appalachian Region. Mining Engineering *62*: 44–49.

36. Scarisbrick DA, Quinlan TR [2002]. Health surveillance for coal workers' pneumoconiosis in the United Kingdom 1998–2000. Ann Occup Hyg *46* (Suppl. 1):254–256.

37. Brief RS, Scala RA [1975]. Occupa-tional exposure limits for novel work schedules. Am Ind Hyg Assoc J *36*:467–469.

38. Kenny LC, Hurley F, Warren ND [2002]. Estimating the risk of contracting pneumoconiosis in the UK coal mining industry. Ann Occup Hyg *46*(Suppl. 1):257–260.

39. Boden LI, Gold M [1984]. The accuracy of self-reported regulatory data: The case of coal mine dust. Am J Ind Med *6*:-440.

40. U.S. Mine Safety and Health Administration [1993]. Report of the Statistical Task Team of the Coal Mine Respirable Dust Task Group. Washington DC: US Department of Labor,

41. CDC [2004]. Changing patterns of pneumoconiosis mortality—United States, 1968-2000. MMWR Morb Mortal Wkly Rep *53*:627–632.

42. CDC [2009]. Coal workers' pneumoconiosis-related years of potential life lost before age 65 years—United States, 1968-2006. MMWR Morb Mortal Wkly Rep *58*:1412–1416.

43. Starzynski Z, Marek K, Kujawska A, Szymczak W [1996]. Mortality among coal miners with pneumoconiosis in Poland. Int J Occup Med Environ Health *9*:279–289.

44. Yi Q, Zhang Z [1996]. The survival analyses of 2738 patients with simple pneumoconiosis. Occup Environ Med *53*:129–135.

45. Meijers JM, Swaen GM, Slangen JJ [1997]. Mortality of Dutch coal miners in relation to pneumoconiosis, chronic obstructive pulmonary disease, and lung function. Occup Environ Med *54*:708–713.

46. Attfield MD, Kuempel ED [2008]. Mortality among U.S. underground coal miners: A 23-year follow-up. Am J Ind Med *51*:231–245.

47. Miller BG, MacCalman L [2010]. Cause-specific mortality in British coal workers and exposure to respirable dust and quartz. Occup Environ Med *67*:270–276.

48. Dalal NS, Newman J, Pack D, Leonard S, Vallyathan V [1995]. Hydroxyl radical generation by coal mine dust: Possible implication to coal workers' pneumoconiosis. Free Radic Biol Med *18*:11–20.

49. Wallace WE, Keane MJ, Harrison JC, Stephens JW, Brower PS, Grayson RL, Attfield

MD [1995]. Surface properties of silica in mixed dusts. In: Castranova V, Vallyathan V, Wallace WE eds. Silica and silica-induced lung diseases. Boca Raton: CRC Press, pp. 107–117.

50. Huang X, Li W, Attfield MD, Nadas A, Frenkel K, Finkelman RB [2005]. Mapping and prediction of coal workers' pneumoconiosis with bioavailable iron content in the bituminous coals. Environ Health Perspect *113*:964–968.

51. Zhang Q, Dai J, Ali A, Chen L, Huang X [2002]. Roles of bioavailable iron and calcium in coal dust-induced oxidative stress: possible implications in coal workers' lung disease. Free Radic Res *36*:285–294.

52. McCunney RJ, Morfeld P, Payne S [2009]. What component of coal causes coal workers' pneumoconiosis? J Occup Environ Med *51*:462–471.

53. Kuempel ED, Attfield MD, Vallyathan V, Lapp NL, Hale JM, Smith RJ, Castranova V [2003]. Pulmonary inflammation and crystalline silica in respirable coal mine dust: dose-response. J Biosci *28*:61–69.

54. Page SJ, Organiscak JA [2000]. Suggestion of a cause-and-effect relationship among coal rank, airborne dust, and incidence of workers' pneumoconiosis. AIHAJ *61*:785–787.

55. Soutar CA, Hurley JF, Miller BG, Cowie HA, Buchanan D [2004]. Dust concentrations and respiratory risks in coalminers: key risk estimates from the British Pneumoconiosis Field Research. Occup Environ Med *61*:477–481.

56. Buchanan D, Miller BG, Soutar CA [2005]. Quantitative relations between exposure to respirable quartz and risk of silicosis. Occup Environ Med *60*:159–164.

57. Coggon D, Taylor AN [1998]. Coal mining and chronic obstructive pulmonary disease: a review of the evidence. Thorax *53*:398–407.

58. Henneberger PK, Attfield MD [1996]. Coal mine dust exposure and spirometry in experienced miners. Am J Respir Crit Care Med *153*:1560–1566.

59. Henneberger PK, Attfield MD [1997]. Respiratory symptoms and spirometry in experienced coal miners: Effects of both distant and recent coal mine dust exposures. Am J Ind Med *32*:268–274.

60. Carta P, Aru G, Barbieri MT, Avata-neo G, Casula D [1996]. Dust exposure, respiratory symptoms, and longitudinal decline in lung function in young coal miners. Occup Environ Med *53*:312–319.

61. Beeckman LF, Wang ML, Petsonk EL, Wagner GR [2001]. Rapid declines in FEV1 and subsequent respiratory symptoms, illnesses, and mortality in coal miners in the United States. Am J Respir Crit Care Med *163*:633–639.

62. Naidoo RN, Robins TG, Becklake M, Seixas N, Thompson ML [2007]. Cross-shift peak expiratory flow changes are unassociated with respirable coal dust exposure among South African coal miners. Am J Ind Med *50*:992–998.

63. Naidoo RN, Robins TG, Seixas N, Lalloo UG, Becklake M [2006]. Respirable coal dust exposure and respiratory symptoms in South-African coal miners: a comparison of current and ex-miners. J Occup Environ Med *48*:581–590.

64. Naidoo RN, Robins TG, Seixas N, Lalloo UG, Becklake M [2005]. Differential respirable dust related lung function effects between current and former South African coal

miners. Int Arch Occup Environ Health 78:293–302.

65. Wang ML, Petsonk EL, Beeckman LF, Wagner GR [1999]. Clinically important FEV1 declines among coal miners: an exploration of previously unrecognized determinants. Occup Environ Med 56:837–844.

66. Seixas NS, Robins TG, Attfield MD, Moulton LH [1992]. Exposure-response relationships for coal mine dust and obstructive lung disease following enactment of the Federal Coal Mine Health and Safety Act of 1969. Am J Ind Med 21:715–734.

67. Wang ML, Wu ZE, Du QG, Petsonk EL, Peng KL, Li YD, Li SK, Han GH, Attfield MD [2005]. A prospective cohort study among new Chinese coal miners—The early pattern of lung function change. Occup Environ Med 62:800–805.

68. Wang ML, Wu Z-E, Du Q-G, Peng K-L, Li Y-D, Li S-K, Han G-H, Petsonk EL [2007]. Rapid decline in forced expiratory volume in 1 second (FEV1) and the development of bronchitic symptoms among new Chinese coal miners. J Occup Environ Med 49:1143–1148.

69. Kuempel ED, Stayner LT, Attfield MD, Buncher CR [1995]. Exposure-response analysis of mortality among coal miners in the United States. Am J Ind Med 28:167–184.

70. Sircar K, Hnizdo E, Petsonk E, Attfield M [2007]. Decline in lung function and mortality: implications for medical monitoring. Occup Environ Med 64:461–466.

71. Naidoo RN, Robins TG, Murray J [2005]. Respiratory outcomes among South African coal miners at autopsy. Am J Ind Med 48:217–224.

72. Vallyathan V, Green FHY, Brower P, Attfield M [1997]. The role of coal mine dust exposure in the development of pulmonary emphysema. Ann Occup Hyg 41:352–357.

73. Kuempel ED, Wheeler MW, Smith RJ, Vallyathan V, Green FH [2009]. Contributions of dust exposure and cigarette smoking to emphysema severity in coal miners in the United States. Am J Respir Crit Care Med 180:257–264.

74. Kuempel ED, Vallyathan V, Green FHY [2009]. Emphysema and pulmonary impairment in coal miners: Quantitative relationship with dust exposure and cigarette smoking. Journal of Physics Conference Series 151:1–8.

75. Brichet A, Wallaert B, Gosselin B, Remy-Jardin M, Voisin C, Lafitte JJ, Tonnel AB [1997]. Primary diffuse interstitial fibrosis in coal miners: a new entity? Rev Mal Respir 14:277–285.

76. Brichet A, Tonnel AB, Brambilla E, Devouassoux G, Remy-Jardin M, Copin MC, Wallaert B [2002]. Chronic interstitial pneumonia with honeycombing in coal workers. Sarcoidosis Vasc Diffuse Lung Dis 19:211–219.

77. International Agency for Research on Cancer. [1997]. IARC monographs on the evaluation of carcinogenic risks to humans: silica, some silicates, coal dust, and para-aramid fibrils, pp. 337–406. Geneva, Switzerland, World Health Organization, International Agency for Research on Cancer.

78. Battelli LA, Ghanem MM, Kashon ML, Barger M, Ma JY, Simokevitz RL, Miles PR, Hubbs AF [2008]. Crystalline silica is a negative modifier of pulmonary cytochrome P-4501A1 induction. J Toxicol Environ Health 71:521–532.

79. Ghanem MM, Batteli LA, Mercer RR, Scabilloni JF, Kashon ML, Ma JY, Nath J, Hubbs AF [2006]. Apoptosis and Bax expression are increased by coal dust in the polycyclic aromatic hydrocarbon-exposed lung. Env Health Perspect 114:1367–1373.

80. Ghanem MM, Porter D, Batteli LA, Vallyathan V, Kashon ML, Ma JY, Barger MW, Nath J, Castranova V, Hubbs AF [2004]. Respirable coal dust particles modify cytochrome P4501A1 (CYP1A1) expression in rat alveolar cells. Am J Respir Cell Mol Biol 31:171–183.

81. Page SJ, Volkwein JC, Vinson RP, Joy GJ, Mischler SE, Tuchman DP, McWilliams LJ [2008]. Equivalency of a personal dust monitor to the current United States coal mine respirable dust sampler. J Environ Monit 10:96–101.

82. U.S. Department of Labor [2010]. 30 CFR Part 74 RIN 1219-AB61 Coal Mine Dust Sampling Devices. Federal Register 75:17512–17529.

83. Weeks JL [2003]. The fox guarding the chicken coop: Monitoring exposure to respirable coal mine dust, 1969–2000. Am J Public Health 93:1236–1244.

84. Weeks JL [2006]. The Mine Safety and Health Administration's criterion threshold value policy increases miners' risk of pneumoconiosis. Am J Ind Med 49:492–498.

85. Love RG, Miller BG, Groat SK, Hagen S, Cowie HA, Johnston PP, Hutchison PA, Soutar CA [1997]. Respiratory health effects of opencast coalmining: a cross sectional study of current workers. Occup Environ Med 54:416–423.

www.ingramcontent.com/pod-product-compliance
Lightning Source LLC
Chambersburg PA
CBHW081905170526
45167CB00007B/3151